Doppler Ultrasound
in the Diagnosis of
Cerebrovascular
Disease

ULTRASOUND IN BIOMEDICINE RESEARCH SERIES

Series Editor: **Professor D. N. White,** *Queen's University, Kingston, Canada*

Doppler Ultrasound in the Diagnosis of Cerebrovascular Disease

Edited by

Professor Robert S. Reneman, M.D.

and

Arnold P. G. Hoeks, EE, M.Sc.

Departments of Physiology and Biophysics, Biomedical Centre, University of Limburg, Maastricht, The Netherlands

RESEARCH STUDIES PRESS
A DIVISION OF JOHN WILEY & SONS LTD.
Chichester · New York · Brisbane · Toronto · Singapore

RESEARCH STUDIES PRESS

Editorial Office:
58B Station Road, Letchworth, Herts. SG6 3BE, England

Library of Congress Cataloging in Publication Data:
Main entry under title:

Doppler ultrasound in the diagnosis of cerebrovascular
   disease.
   (Ultrasound in biomedicine research series; v.5)
   Includes index.
   1. Cerebrovascular disease—Diagnosis.
   2. Diagnosis, Ultrasonic.   3. Doppler effect.
   I. Reneman, Robert S.   II. Hoeks, Arnold P. G.
   III. Series.
   RC388.5.D64   616.8'107543   81-19854
ISBN 0 471 10165 6   AACR2

British Library Cataloguing in Publication Data:

Doppler ultrasound in the diagnosis of
   cerebrovascular disease.—(Ultrasound in biomedicine
   research series; v.5)
   1. Cerebrovascular diseases—Diagnosis
   2. Diagnosis, Ultrasonic
   I. Reneman, Robert S.   II. Hoeks, Arnold P. G.
   III. Series
   616.8'107543   RC388.5

   ISBN 0 471 10165 6

Printed in Great Britain

Preface

In the past decade several non-invasive techniques have been developed to evaluate the cerebrovascular circulation. Without ignoring the potential of such techniques as ocular pneumoplethysmography and carotid phonoangiography, the present book confines itself to ultrasonic techniques. The main reason for this is that Doppler ultrasound techniques enjoy some special advantages in diagnosing cerebrovascular disease. Firstly, with Doppler techniques local information - that is at various sites along and in an artery - can be obtained, at least when the artery is accessible to ultrasound. Secondly, disturbances in the flow pattern or more specifically in the velocity profile, which are known to occur with low grade stenosis of an artery, can be detected with these techniques. Thirdly, Doppler devices can be used indirectly to evaluate the functional state of the cerebral circulation. Beside these advantages, ultrasound can be used to image vessel wall structures (B-mode imaging) and the circulation (Doppler velocity imaging). These imaging techniques are especially suitable to localize the site of sampling in relation to a vascular lesion, facilitating the interpretation of recorded signals. Lastly, Doppler investigations are atraumatic and safe and hence especially suitable for epidemiological and follow-up studies.

Although it is the aim of this book to inform clinicians about the various Doppler ultrasound methods currently in use to diagnose cerebrovascular disease, we found it appropriate to start the book with chapters on the anatomy and physiology of the cerebral circulation (Chapter 1) as well as on basic fluid dynamics (Chapter 2) and some physical aspects (Chapter 3) of the (cerebral) arterial circulation. Better understanding of the anatomy and physiology may facilitate the interpretation of the described diagnostic procedures. Insight into basic fluid dynamics and some physical aspects is essential for appreciation of the variables and parameters used by the various authors especially in the detection of flow disturbances induced by vascular lesions. In a separate chapter (Chapter 4) atten-

tion is paid to the technical aspects of Doppler instruments. In
this chapter the most common techniques, using either continuous
wave (CW) or pulsed emission, are discussed.

In some of the more fundamental chapters the use of formulas
could not be avoided. Since physicians are not generally acquainted
with the use of these mathematical relations, a compromise was found
by modelling the content of these chapters in such a way that one can
easily skip the formulas and still retain the scope of the chapter.
In these chapters the symbols used are listed after the references.

The techniques described to diagnose cerebrovascular lesions com-
prise hand-held probing, using CW Doppler and analog velocity waveforms (Chapters 5 and 6), velocity imaging and audiospectrum analysis,
using CW Doppler instruments (Chapters 7 and 8) as well as pulsed
Doppler systems combined with B-mode imaging (Chapter 9). The last
chapter discusses recent ultrasonic developments that may be of
clinical importance. In this chapter an attempt is also made to indicate the potentialities of the instruments available, which may
facilitate their selection for clinical use.

Although at the present state of the art most of the described
techniques can accurately detect vascular lesions associated with
substantial narrowing of the carotid artery and can distinguish high
grade stenosis from total occlusion, it should be noted that not all
of these techniques have been evaluated in terms of their diagnostic
accuracy, sensitivity or specificity. Accurate diagnosis of vascular
lesions without substantial narrowing of the carotid arteries is
still problematic, although promising results have been obtained
both with CW (Chapter 8) and pulsed Doppler systems (Chapter 9). The
detection of these lesions is important because they are considered
to be a source of emboli.

September 1981   Robert S. Reneman

                                        Arnold P.G. Hoeks

Acknowledgements

We wish to express our thanks to the various authors for accepting without hesitation our invitation to participate in this book. We are grateful to Dr. Denis White for his valuable advice and his help in correcting the manuscripts of the authors whose native tongue is not English. But most of all we would like to thank Mrs. Mariet de Groot and Mrs. Joke Hoozemans-Koreman for carrying the heavy secretarial load with fortitude as well as for their help in preparing the manuscripts and typing them in their final form. A multi-author book is not only a pain in the neck for the editors but also for their secretaries, especially when camera-ready copies of the manuscripts have to be produced. Without their support this book would have looked different.

Contributing Authors

M.G.J. Arts, Ph.D.
Department of Biophysics
Biomedical Center
University of Limburg
P.O. Box 616
6200 MD Maastricht/The Netherlands

M.G. Beasley, B.Sc.
Ultrasonic Angiology Unit
Guy's Hospital
London Bridge SE1 9RT/England

M.E.H. Van Dongen, Ph.D.
Department of Physics
University of Technology
P.O. Box 513
5600 MB Eindhoven/The Netherlands

G. Fell, F.R.A.C.S.
Assistant Vascular Surgeon
Austin Hospital
Heidelberg 3084/Australia

R.G. Gosling, Ph.D., F.Inst. P.
Reader in Physics Applied to Medicine
Head of Ultrasonic Angiology Unit
Clinical Science Laboratories, Floor 17
Guy's Hospital Tower
London Bridge SE1 9RT/England

A.P.G. Hoeks, EE, M.Sc.
Department of Biophysics
Biomedical Center
University of Limburg
P.O. Box 616
6200 MD Maastricht/The Netherlands

J. Jonkman, M.D.
Head Research Unit for Clinical Neurophysiology T.N.O.
Westeinde Hospital
Lijnbaan 32
2512 VA The Hague/The Netherlands

R.R. Lewis, M.B., M.R.C.P.
Consulting Physician
Ultrasonic Angiology Unit
Guy's Hospital
London Bridge SE1 9RT/England

J.M.F. Mol, M.D.
Professor of Clinical Neurophysiology
Department of Clinical Neurophysiology
University of Limburg
P.O. Box 616
6200 MD Maastricht/The Netherlands

P.C.M. Mosmans, M.D.
Research Unit for Clinical Neurophysiology T.N.O.
Westeinde Hospital
Lijnbaan 32
2512 VA The Hague/The Netherlands

L. Pourcelot, Ph.D., M.D.
Professor of Biophysics
Head of the Department of Nuclear Medicine and Ultrasound
Medical Facultay
François Rabelais University
2 bis Boulevard Tonnelé
37032 Tours Cedex/France

R.S. Reneman, M.D.
Professor of Physiology
Department of Physiology
Biomedical Center
University of Limburg
P.O. Box 616
6200 MD Maastricht/The Netherlands

M.P. Spencer, M.D.
Director Institute of Applied Physiology and Medicine
701 16th Avenue
Seattle, WA 98122/USA

A.A. Van Steenhoven, Ph.D.
Department of Mechanical Engineering
University of Technology
P.O. Box 513
5600 MB Eindhoven/The Netherlands

D.E. Strandness, M.D.
Professor of Surgery
Department of Surgery
University of Washington
Medical School
Seattle, WA 98195/USA

Table of Contents

CHAPTER 1
Basic Anatomy, Physiology and Pathology of Human Cerebral Circulation

E. J. Jonkman *and* P. C. M. Mosmans

1.1 INTRODUCTION

This chapter does not pretend to give an exhaustive survey of the anatomy, physiology and pathology of the cerebral circulation. This short introduction is aimed especially at those readers who are interested in Doppler flow velocity measurements and have more technical than medical training. The medically trained may find it a useful summary.

The literature about the anatomy and physiology of the cerebral circulation is quite extensive. Descriptions of the normal cerebral vascular supply and its anatomical variations date as far back as the 17th century. Anatomical descriptions of the cerebral circulation can be found in many textbooks in general use. The study of the physiology of the cerebral circulation started in the latter half of the 19th century. The possibility of studying human cerebral circulation in vivo by the use of radioactive isotopes has been an advance which has led to a large number of publications in this field. The studies of the physiology and pathology of the cerebral circulation have been published in many articles from numerous research groups. The reports are often overlapping in subject and results. Discrepancies in results and interpretation are frequently encountered. Due to the complexity of cerebral circulation measurements, observations in normal volunteers have been rather scarce until now.

Our personal views of the subject only are presented in this chapter in order not to burden the reader with confusing discrepancies which arise when all the relevant literature is taken into consideration. This implies that a complete list of relevant literature will not be given and only a list of "recommended reading" is given at the end of this chapter. The following paragraphs deal, in sequential order, with anatomy (including normal variants), physiology and pathology. This division may sometimes appear to be arbitrary and overlapping.

Relevant Doppler problems and possibilities will be mentioned briefly; for details we may refer to the following chapters in this book.

1.2   ANATOMY OF THE CEREBRAL CIRCULATION
1.2.1 Introduction

Only those anatomical details will be described here that are relevant for the understanding of Doppler flow measurements. (Figs. 1.3, 1.4 and 1.5 were drawn from arteriograms in order to reproduce the normal anatomical relations as much as possible).

1.2.2 Aorta And Arteries Of The Neck

The branches of the aortic arch which lies just behind the sternum are in sequential order from right to left: innominate artery (truncus brachiocephalicus), left common carotid artery and left subclavian artery (Figs. 1.1 and 1.2). The innominate artery bifurcates into the right common carotid and right subclavian arteries. Both common carotid arteries pass up from the thorax on each side of the neck and each divides into the external and internal carotid arteries. The internal carotid artery runs up to the base of the skull without branching and supplies the main part of the blood flow for the brain. The external carotid artery has many branches supplying the blood for the organs of the neck, the muscles and other tissues on the outside of the skull, the skull and the dura mater (one of the three membranes inside the skull surrounding the brain). In the neck region (Fig. 1.2) we find between both carotid arteries the trachea, larynx and thyroid gland. Both arteries are situated a little behind and on the outside of the thyroid gland. For Doppler studies it is important to remember that an enlargement of the thyroid gland may induce a displacement of the carotid arteries in the posterio-lateral direction. Moreover, the lower part of the common carotid artery is covered by the sternocleidomastoid muscle.

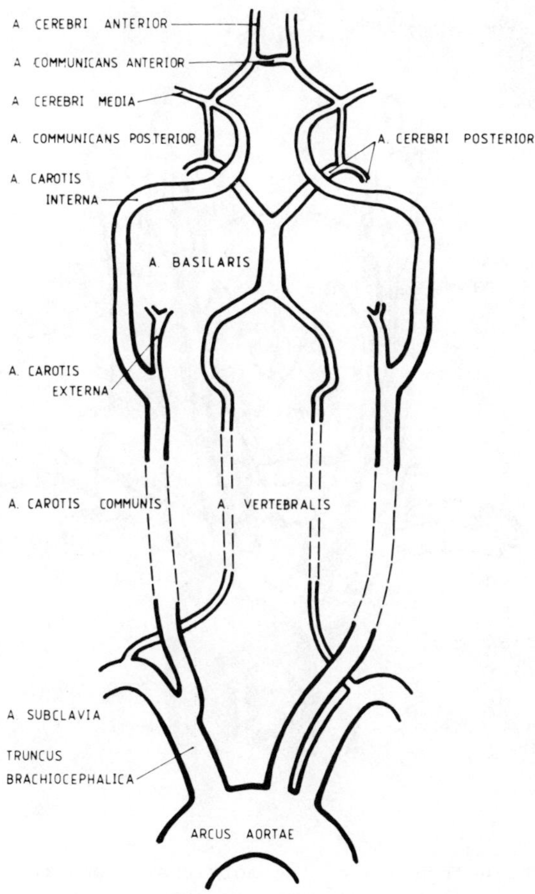

FIG. 1.1   Diagram of the neck vessels and circle of Willis.

In most people the internal carotid artery runs a little bit to the outside and behind the external carotid artery. The bifurcation of the common carotid artery can vary considerably in level (Fig. 1.3). In Doppler examinations one should be aware of the possibility that very low bifurcations (below the 6th cervical vertebra) do occur. The point of bifurcation tends to be symmetrical.

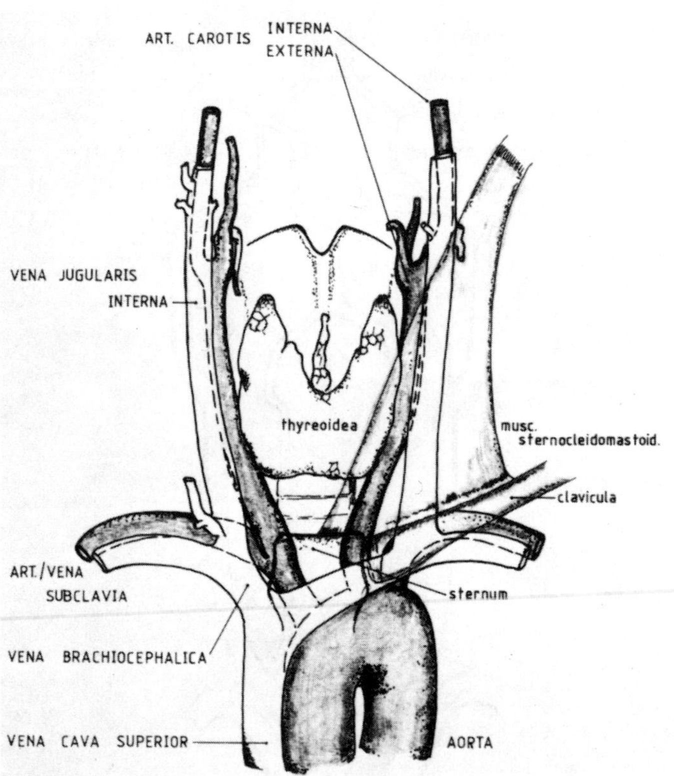

FIG. 1.2   Main branches of the aortic arch and the anatomical relation between the carotid arteries and surrounding structures.

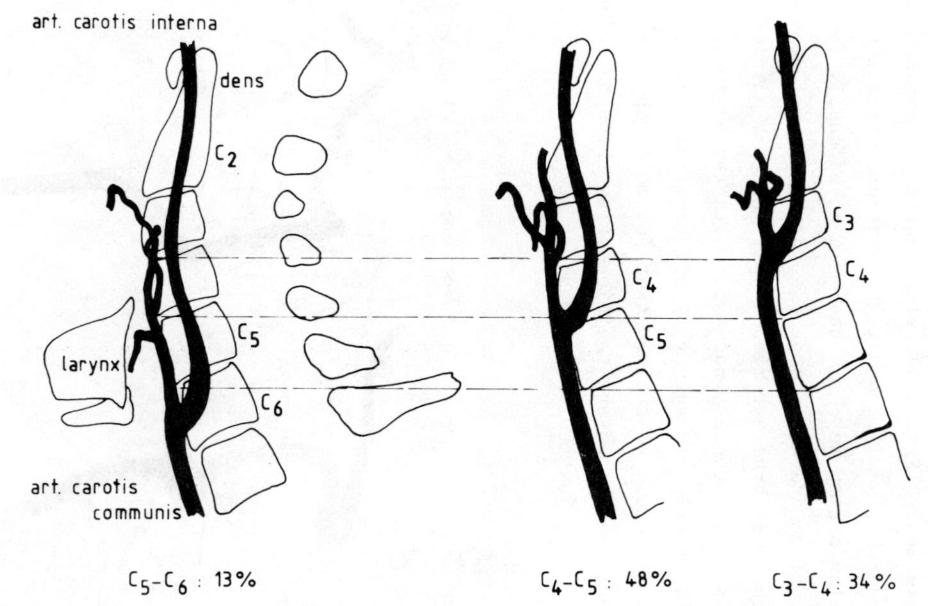

FIG. 1.3   Incidence of the different levels of the carotid bifurcation.

1.2.3   The Vertebral Arteries

The left and right vertebral arteries are branches from the left and
right subclavian arteries respectively. The arteries ascend to the
foramen transversarium of the 6th cervical vertebra and having passed
through that foramen and those of the next succeeding cervical verte-
brae, form some loops (Fig. 1.4) before entering the skull through
the foramen magnum. The left and right vertebral arteries join to
form the basilar artery which is the main blood supply for the brain
stem and cerebellum and a minor part of both cerebral hemispheres.
Branches of the vertebral arteries supply the neck muscles and the
cervical spinal cord. Connections exist with the branches of the external carotid arteries (a survey of all cervical vessels is given
in Fig. 1.5). It should be noticed that for most Doppler apparatus
it is only possible to measure the blood flow through the vertebral
arteries at a level just below the base of the skull. The loops of

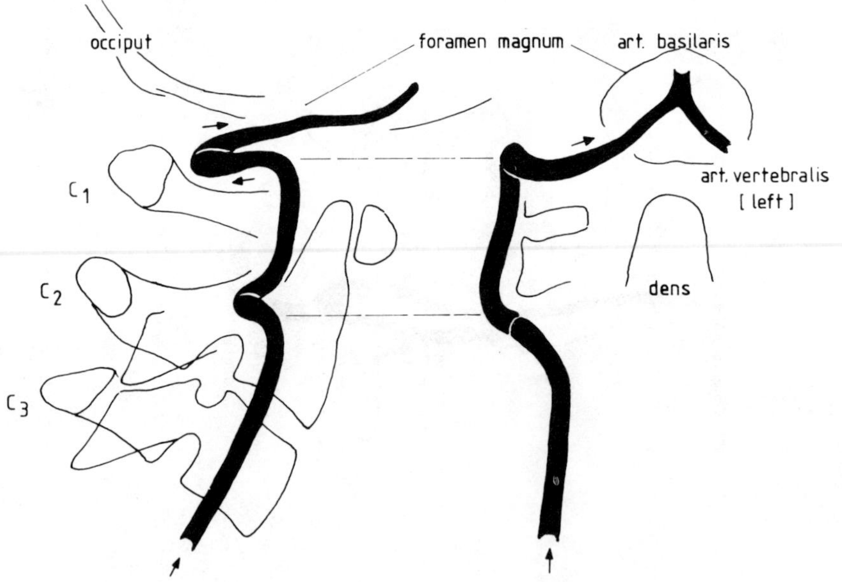

FIG. 1.4   The loops of the vertebral artery as seen in lateral
  view (left) and antero-posterior view (right).

the vertebral artery in this region make it rather difficult to
determine from the Doppler signal whether the flow is in the normal
direction or reversed.

1.2.4   Intracranial Vessels

The basilar artery branches into the left and right posterior cerebral arteries which are connected with both internal carotid arteries. Since there is also a connection between the anterior branches
of both internal carotid arteries, a circular arterial system is
formed known as the circle of Willis. The whole circle consists of
the two posterior cerebral arteries, two posterior communicating
arteries, two anterior cerebral arteries and one anterior communicating artery (Fig. 1.1). From the circle the main arteries supplying
the cerebral hemispheres arise: anterior cerebral arteries, the
middle cerebral arteries and the posterior cerebral arteries. Of
special interest is the ophthalmic artery, supplying the orbit and
its contents. This artery is a small branch from the internal carotid
artery in its intracranial segment. Two of its branches, the supraorbital and frontal arteries leave the orbit and are connected with
small branches of the external carotid artery (Fig. 1.6). This connection is the only significant physiological connection between the
external and internal carotid artery system (see section 1.2.6).

1.2.5   Variations

In the branching of the aortic arch several variations are known to
exist. In some people the right common carotid artery and right subclavian artery are separate branches from the aortic arch which implies the absence of an innominate artery. In other cases an innominate artery exists on the right side as well as on the left side.
The origin of the vertebral arteries is not always found in the subclavian arteries, sometimes the vertebral arteries are primary
branches of the aortic arch. The variations at the level of the
carotid bifurcation have been mentioned above. In the majority of
all people the internal carotid artery lies on the outside and poste-

rior of the external carotid artery. In about 9% however the situation is reversed: the internal carotid artery lying on the inside of the external carotid artery. Specific abnormalities of the internal carotid artery are so called "coiling", "tortuosity" and "kinking".

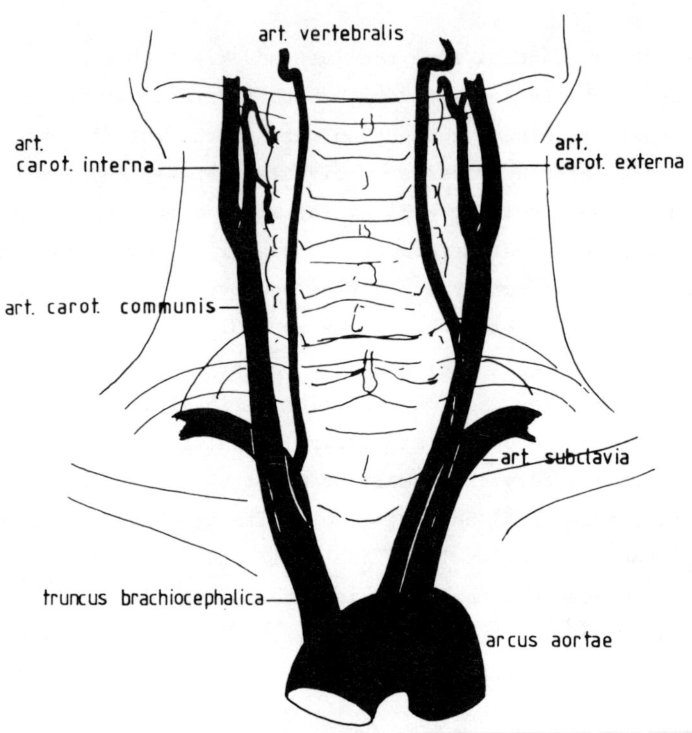

FIG. 1.5   Large arteries of the neck as seen in
    a normal arteriogram.

According to Fields and co-investigators (1965) coiling and tortuosity are to be considered as congenital variants which can be accentuated in patients with arteriosclerosis and/or hypertension. Tortuosity can be described as the presence of "S" or "C" like loops. This phenomenon is more clearcut in coiling (even circular loops do occur). Arteriosclerosis is probably the main cause in the origin of kinking. The incidence of these phenomena being (Weibel and Fields, 1965) tortuosity 15%, coiling 3%, kinking 0.5%. In 45% the left vertebral artery is larger and in 21% the right.

The mean diameter of the internal carotid artery at the level of the
atlas is 5.90 mm $\pm$ 0.73 mm (Gabrielsen and Greitz, 1970). Numerous
other variants do occur with a rather low incidence: aplasia or hypoplasia of the internal carotid arteries, abnormal connections between
carotid and vertebral system etc.

In many cases the circle of Willis, as described above, is not
fully developed. Fields and collaborators (1965) concluded from a
study of literature that in 59% of the cases the circle did not show
the "ideal" configuration. The anomalies in the circle of Willis are
of clinical interest. The underdevelopment of a part of the circle
of Willis can mean the lack of an important anastomosis when one of
the main supplying arteries is occluded.

1.2.6   Anastomosis

Several connections exist between the arterial systems described
above. The most important of these anastomoses are:
- several branches of the external carotid artery are connected with
  branches of the ophthalmic artery thus forming an anastomosis
  between the external and internal carotid arteries (Fig. 1.6, nr 1)
- branches of the external carotid artery are connected with branches
  of the vertebral artery (Fig. 1.6, nr 2)
- the circle of Willis forms the main anastomosis between the internal
  carotid arteries (and their main branches) and the vertebral arteries (Fig. 1.6, nr 3)
- the so called "leptomeningeal anastomosis" in which interconnections occur between the cortical branches of the anterior, middle
  and posterior cerebral arteries.

Anatomically and angiographically the presence of other anastomoses
has been established. However, it is doubtful that these anastomoses
can have a real function in cases of vessel occlusion.

1.2.7   The Venous System

The drainage of the blood from the brain mainly takes place via the
cortical veins into the dural sinus and hence into the internal

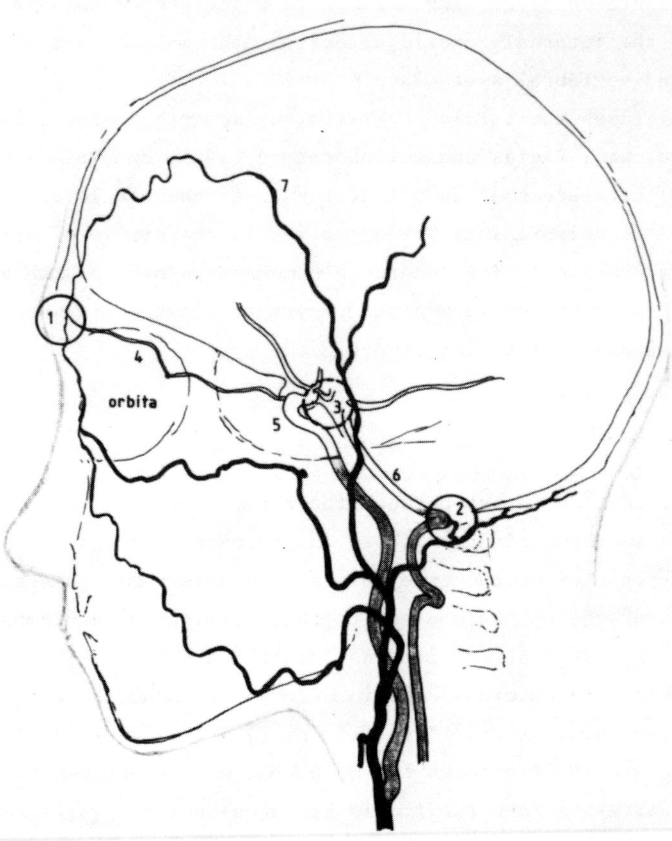

FIG. 1.6   Main anastomoses:
(1) connection between ophthalmic artery and branches of ex-
    ternal carotid artery
(2) anastomosis between vertebral artery and external carotid
    artery
(3) circle of Willis
        Main arteries:
(4) ophthalmic artery
(5) internal carotid artery
(6) basilar artery
(7) superficial temporal artery (branch of external carotid
    artery)

jugular veins and the venous plexus around the spinal cord. The blood
of the skull and the face is collected in the left and right external
jugular veins which join the internal jugular vein. The anatomy of
the venous system shows many variations. For Doppler measurements it
is important that the largest veins are situated in front and lateral
to the common carotid artery (Fig. 1.2).

1.3   PHYSIOLOGY OF THE CEREBRAL CIRCULATION
1.3.1 <u>Introduction</u>

Compared with other tissues the brain has remarkably high metabolic
needs. Brain metabolism is only slightly enhanced when cerebral activation is present. Recently it appeared possible to demonstrate small
differences in blood flow during wakefulness compared with sleep and
during activation compared with rest. These differences are relatively small. To illustrate the metabolic need of the brain: cerebral
glucose utilisation is about 70% of the resting glucose output
(while the brain comprises only 2% of the average body weight). The
normal average blood flow through the brain tissue is approximately
60 ml/min / 100 gram. For the average brain weight this amounts to
900-1100 ml/min, which is 12-15% of the resting cardiac output. The
cerebral blood flow value mentioned above (60 ml/min/100 gram) is
however a mean value. Blood flow through the white matter of the
brain is considerably lower, blood flow through the grey matter considerably higher. Moreover regional differences are also always
present.

There are two differences between the brain circulation and the
circulation through other parts of the body. First of all, the brain
tissue is extremely dependent on a continuous and adequate supply
of oxygen and glucose. In animal experiments the upper limit of circulatory arrest which can be survived by the brain tissue is 15 minutes. However, survival of the neurons only takes place when, immediately after this ischemic period, optimal conditions (blood pressure, respiration etc.) are reinstituted. In clinical practice this
is rarely achieved. The period of circulatory arrest which leads to
permanent brain damage is therefore much shorter in man: about 5-7

minutes. Secondly, enhancement or diminution of the cerebral circulation does not, within certain limits, produce subjective or objective alterations in the clinical state; but if circulation falls
below a certain level (20 ml/min/100 gram) reversible impairment
of cerebral function occurs; below another level (10 ml/min / 100 gram)
irreversible brain damage is certain to occur. These values are only
approximations because it will be apparent that experiments can be
done in animal studies only. The total blood flow reaching the brain
is divided between the two carotid arteries and the two vertebral
arteries. Even in normals asymmetries occur (at least up to 30%) in
the flow through both carotid arteries and sometimes these asymmetries are even larger for the flow through both vertebral arteries.
Moreover, the watershed between the area of the carotid arteries and
the vertebral arteries can differ from person to person. This implies
that if one tries to measure the total cerebral blood flow (e.g. by
ultrasound) one always has to measure all four vessels.

1.3.2 Cerebral Blood Flow And Blood Pressure

The cerebral blood flow through the brain is determined by
the pressure gradient between input and output of the system and the
cerebrovascular resistance (Fig. 1.7). In animal experiments no
essential differences could be found between pulsating flow and non-
pulsating flow. It can be assumed that in man a continuous (non-
pulsating) flow is well tolerated (extracorporeal circulation in
open-heart surgery) and is subject to the same regulating mechanisms
existing for the pulsating flow (the term "cerebrovascular resistance" instead of perhaps more appropriate "cerebrovascular impedance" is still in general use). The pressure at the output side of
the system is in man rather low, the normal venous pressure in the
skull about equals the tissue pressure (= intracranial pressure).
The values show oscillations with the respiration and the heart rate
but the normal values (1-2 kPa) are very low compared with the input
pressure. The venous system draining the blood off the brain has so
many interconnections that even an occlusion of one of the major
veins (vv. jugulares) does not lead to an increase in output

(= venous) pressure in the brain. There are also no clinical neurological syndromes known to exist caused by extracerebral venous occlusion. Intracranially there are parts of the venous system where
anastomoses are scarce or even absent. An occlusion of the venous
system in such a vulnerable spot may lead to a severe impairment of
local cerebral perfusion and be the cause of severe dysfunction.

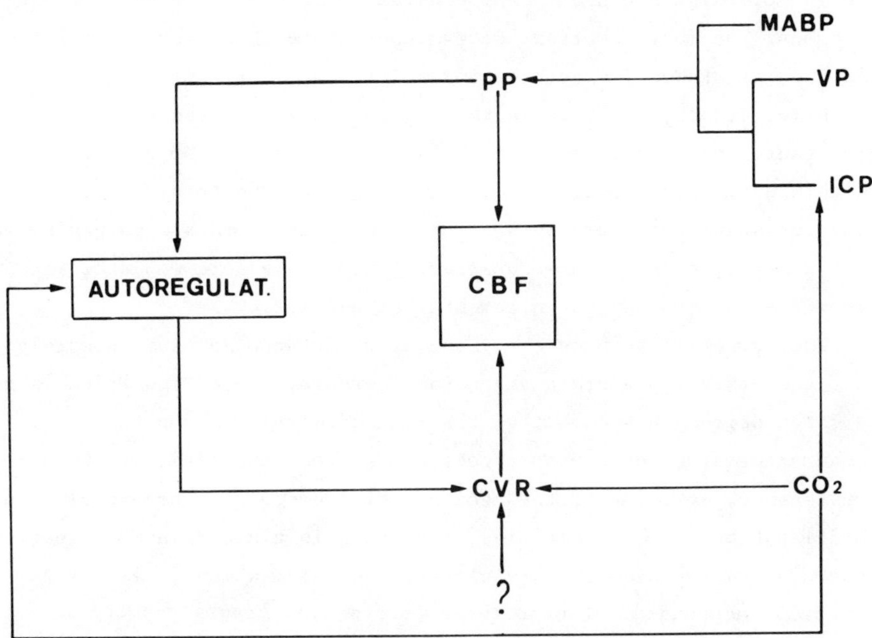

FIG. 1.7   Relation between mean arterial blood pressure (MABP),
venous pressure (VP), intracranial pressure (ICP), arterial P_{CO_2},
vascular resistance (CVR), cerebral blood flow (CBF) and autoregulation.

So in normals and most of our patients, the pressure gradient
(= perfusion pressure) is determined mainly by the input pressure
(= mean arterial blood pressure) which in man can vary between 9.3
and 22.7 kPa without giving rise to direct neurological symptoms.

The cerebrovascular resistance is partially determined by the
resistance in the capillaries. This resistance can be assumed to be
a constant because in the brain there are no possibilities to shut
off or open important parts of the capillary system. Such a system
could not exist because of the constant high metabolic need of the
neurons. The most important determinant of cerebrovascular resistance
is the resistance in the arterioles (pre-capillary small arteries).
This resistance, in series with the capillary resistance, is not a
constant. The arterioles have the ability to narrow or to dilate
within a certain range. The effect being that the total cerebrovascular resistance can vary between certain limits. If this mechanism was
not present, changes in mean arterial blood pressure would be linearly reflected in changes in cerebral blood flow.

The variability in cerebrovascular resistance makes it possible
for the brain to maintain a constant cerebral blood flow which is to
a large degree independent of the mean arterial blood pressure: if
mean arterial blood pressure goes up, so does the cerebrovascular
resistance, lowering of the mean arterial blood pressure results in
a dilatation of the arterioles, resulting in a decrease in cerebrovascular resistance. This regulatory mechanism which is almost completely independent of structures outside the cranial cavity is
called autoregulation. Autoregulation is a feature present in man and
other mammals, it plays the main part in maintaining a constant cerebral blood flow but can be easily disturbed by cerebral or generalized disorders. Autoregulation however is limited. There is a maximum
possible dilation of the arterioles and also the narrowing of these
microvessels has an obvious limit. In other words: the range of variations of cerebrovascular resistance has an upper and lower limit.
This means that between a certain range (9.3-22.7 kPa) blood flow
through the brain is independent of variations in mean arterial blood
pressure. Outside these limits the blood flow is passively pressure
dependent. A decrease of mean arterial blood pressure below 9.3 kPa

often results in a decrease in cerebral blood flow, an increase over 22.7 kPa mostly results in an increase of cerebral blood flow ("break through" phenomenon). It will be clear that surpassing the limits of autoregulation will be hazardous for the patient: diminishing the blood pressure under the lower limit will endanger the necessary blood supply of the brain, increasing the blood pressure over the upper limit of autoregulation can result in an overstretching of the capillary wall, resulting in brain edema. Although autoregulation is shown to exist in many animals and is undoubtedly present in man, observations in humans are scarce because the limits of autoregulation cannot be tested without surpassing ethical limits.

There are 3 hypotheses for the underlying mechanism of autoregulation:

- the neurogenic theory. The arteries in the brain are supplied by small nerve-fibers, probably originating in the upper brainstem. The importance of these fibers is not completely clear. It does not seem probable that the widening or narrowing of the arterioles necessary for the autoregulation is determined by these fibres only.
- the myogenic theory. The smooth muscle fibres in the arteriolar wall react directly to transmural pressure changes (Bayliss described this effect already in 1902). When pressure in an artery suddenly rises the vessel constricts, when the pressure falls the arteriole dilates. Since these observations were made in vitro a metabolic or neurogenic influence could be excluded.
- the metabolic theory. According to this theory changes in blood flow induced by changes in perfusion pressure are immediately followed by changes in metabolism and therefore by changes of brain tissue pH. A decrease in pH produces a dilatation of the arterioles, an increase in pH a constriction. It is very likely that the three mechanisms described above are working together in autoregulation.

It should be pointed out that an increase in blood pressure does not necessarily have an influence on the diameters of larger arteries. This results in a constancy of the flow velocity e.g. in the carotid artery during modest increases or decreases of mean arterial blood pressure (the cerebral blood flow and the vessel diameter both being maintained at a fixed level).

1.3.3   Influence Of Arterial P_{CO_2} On Cerebral Blood Flow

Hypercapnia causes a dilatation of the cerebral arterioles while hypocapnia produces a cerebral vasoconstriction. There exists an almost linear relationship between changes in arterial P_{CO_2} and cerebral blood flow (1% change in arterial P_{CO_2} giving a change of about 2.2% of the resting value in cerebral blood flow). The physiological meaning might be that acidosis caused by insufficient metabolism gives a vasodilatation and a better perfusion of the tissue. This seems to be the same mechanism as described under "metabolic theory of autoregulation". In brain disease, however, autoregulation can be impaired without disturbances of the CO_2 response or vice versa.

The influence of arterial P_{CO_2} can be important in Doppler flow velocity measurements. Changes in arterial P_{CO_2} result in changes of cerebral blood flow but do not influence the diameter of the larger arteries. An increase in arterial P_{CO_2} results in an increase in the flow velocity in the large vessels leading to the brain, this was demonstrated in a series of young adults (Fig. 1.8).

1.3.4   Influence Of Activation

It was mentioned that the total cerebral blood flow shows only slight variations when an alert state is compared with a resting state. There are, however, studies (Olesen, 1971) showing that an important local increase in cerebral blood flow can be obtained by activating the corresponding cortical region in a physiological way (e.g. voluntary hand movements give a rise in blood flow in the motor region). Also pathological hyperactivity (e.g. focal epilepsy) can produce a focal increase in cerebral blood flow. Because the activated area involved is always only a minor portion of the total brain volume, such increases cannot be detected when total cerebral blood flow or blood flow velocity in the large arteries is studied.

1.4   PATHOPHYSIOLOGY

1.4.1   Loss Of Autoregulation

Autoregulation is present in normals but can be changed or abolished

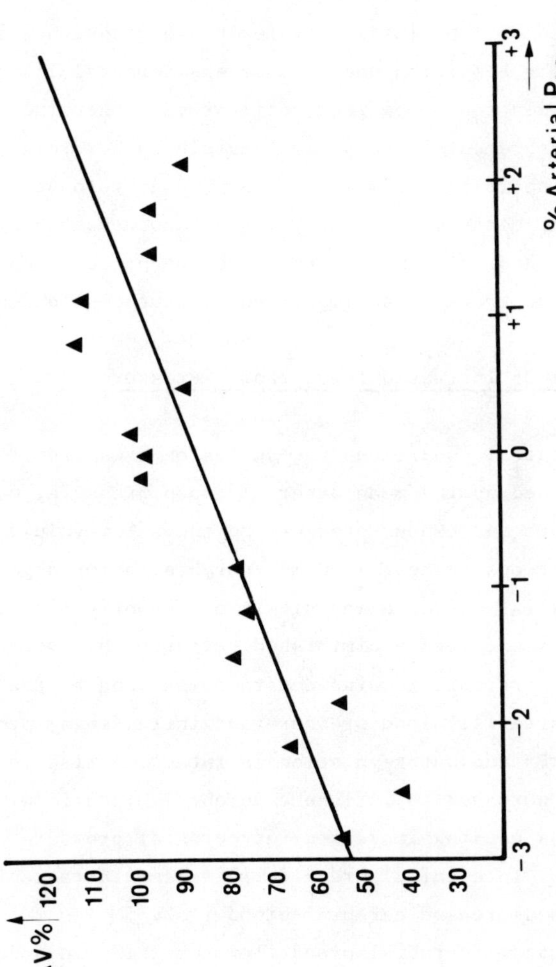

FIG. 1.8   Relation between percentage arterial P_{CO_2} ($\bigcirc$ = resting value) and blood flow velocity (V) measured in the mid-diastolic period (100% = resting value).

by cerebral or generalized disease. In the literature we can find references indicating that autoregulation can be lost globally or focally in patients with cerebral tumours, ischemic cerebral disease, cerebral trauma, meningitis, encephalitis, subarachnoidal bleeding, hydrocephalus, during induced seizures, under hypercapnia, in Shy-Drager syndrome and in diabetes. Moreover, the range of the auto-regulation (9.3–22.7 kPa) can be shifted to higher blood pressure levels in cases of long lasting systemic hypertension. Especially in patients in poor condition due to a severe cerebral lesion one should assume autoregulation to be lost until proven otherwise. Repeated blood flow measurements are hardly feasible in severely ill neurological patients so it is difficult to prove a disturbance of autoregulation. Doppler measurements however are innocuous. Changes in flow velocity with changes in mean arterial blood pressure can put the clinician on the track of an impairment of autoregulation.

1.4.2   Effect Of Increased Intracranial Pressure

Under physiological conditions the perfusion pressure of the brain is mainly determined by the mean arterial blood pressure, both the intra-cranial pressure and venous pressure having a low value. Under pathological conditions (cerebral edema, cerebral hemorrhage, tumours etc.) intracranial pressure may slowly or abruptly rise to high values. This rise causes a diminished perfusion pressure when mean arterial blood pressure remains constant. As long as the difference between mean arterial blood pressure and intracranial pressure equals at least 9.3 kPa and autoregulation is intact, a rise in intracranial pressure does not have to influence cerebral blood flow. However, the same affections causing increased intracranial pressure often disturb autoregulation. In clinical practice increased intracranial pressure often leads to decreased cerebral blood flow.

This decrease in cerebral blood flow resulting in a decrease in flow velocity could be demonstrated in 11 patients with hydrocephalus, subarachnoidal bleeding or cerebral tumours in whom a short lasting increase in intracranial pressure was artificially induced or occurred spontaneously (Fig. 1.9).

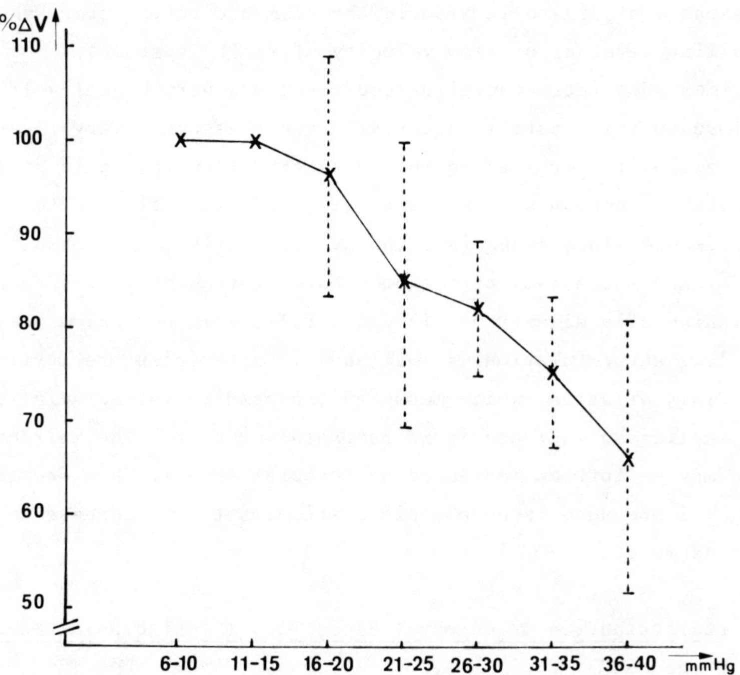

FIG. 1.9   Relation between increase of intracranial
pressure and blood flow velocity measured in the
mid-diastolic period in 11 patients without normal
autoregulation (1 mmHg = 0,133 kPa).

It will be apparent that a disastrous situation arises when the intracranial pressure rises to such heights that the perfusion pressure gets very low and ultimately equals zero. At this point no cerebral circulation is possible any longer and the brain dies.

At this moment there is no flow through the internal carotid artery anymore. In this situation the flow through the common carotid artery mainly represents the flow through the external carotid artery. This situation can be evaluated by Doppler flow velocity measurements (disappearance of diastolic flow in the common carotid artery and a short lasting reversal of flow velocity after the systole).

Sometimes when intracranial pressure reaches very high levels and comes close to the diastolic arterial blood pressure a very interesting reaction is set into action by the body: the systemic arterial blood pressure increases to prevent a complete abolition of the perfusion pressure. This mechanism, the Cushing reflex, is probably one of the defense mechanisms of the body but unfortunately it is usually the beginning of a disastrous vicious circle. When perfusion pressure is very low, autoregulation is lost and all arterioles are maximally dilated. This dilation, which means an increase in intracranial blood volume, implies an increase in intracranial pressure. The intracranial pressure may be further increased by cerebral edema. This increase in intracranial pressure is once again a stimulus to an increase in blood pressure etc.

1.4.3 Redistribution Of Cerebral Blood Flow ("Intracranial Steal Phenomena")

It was already mentioned that hypercapnia (increase in arterial P_{CO_2}) causes dilation of cerebral vessels. This is a useful mechanism in cases with localized increased metabolic needs. However, generalized hypercapnia in patients with localized ischemia sometimes produces a deterioration of the clinical condition rather than an improvement.

The study of the cerebral blood flow in animals with artificially induced local ischemia presented the solution of this problem: the arterioles in the ischemic brain area are already maximally dilated, probably by the fall of tissue pH in the ischemic region. When the vascular bed in the non-ischemic area is dilated by induced hypercapnia, a redistribution of the blood supply takes place in favour of the healthy parts of the brain and at the cost of the already ischemic area. This phenomenon is called "intracerebral steal": the blood is "stolen" from the ischemic area and supplied to the healthy brain tissue. Hypocapnia on the contrary can produce vasoconstriction of vessels in the non-ischemic brain area and has no effect on the maximally dilated arteries in the ischemic area where CO_2 responsiveness is lost. In this situation redistribution of the blood supply is now in favour of the ischemic brain tissue. This is called "inversed steal" or "Robin Hood syndrome" (stealing from the rich and giving to the poor). This mechanism could have clinical implications. Nevertheless the clinical beneficiary effect of hypocapnia (induced by hyperventilation) has never been proved in larger series of patients with ischemic brain disease. It should be pointed out that "intracerebral steal" and "reversed steal" have no influence on the total cerebral blood flow and cannot be detected by flow velocity studies of the large arteries. The same applies to a seldom encountered phenomenon: the so called interhemispheric steal. This is a situation where one hemisphere (e.g. after occlusion of the supplying carotid artery) steals blood from the contralateral hemisphere (e.g. through the circle of Willis). If a partial vascular occlusion is also present in the deprived hemisphere, neurological symptoms may arise originating from the side contralateral to the occlusion of the carotid artery. For problems concerning extracranial steal we may refer to the chapters 4, 5, 6 and 7.

1.4.4   Influence Of Hypoxia

Severe hypoxia (P_{O_2} in the arterial blood <7.3 kPa) produces a dilation of the arterioles. This is an example of an adequate defense mechanism, trying to prevent brain ischemia. The vasodilatation is probably caused by a fall of pH in the ischemic brain tissue. The change from aerobic to anearobic metabolism, where lactate is produced, is responsible for this fall in pH.

1.5   PATHOLOGY OF ISCHEMIC BRAIN DISEASE

1.5.1 Introduction

In the previous section it was pointed out that a diminished mean
arterial blood pressure or an increased intracranial pressure may
lead to an insufficient blood supply to the brain. Also an increase
in the viscosity of the blood may increase the cerebrovascular resistance to the extent that normal flow values cannot be reached. In
clinical practice, however, the three factors mentioned above account
only for a small percentage of the patients with ischemic brain disease. In the majority of cases (focal) ischemic brain disease is
caused by an acquired narrowing of the lumen or occlusion of one or
more of the main cerebral arteries. The main causes for such a partial or total occlusion being either thrombosis, embolism or spasm.

These three affections will be described here briefly. Lastly ischemic brain disease caused by arteriovenous malformations will be
discussed.

1.5.2   Thrombosis In The Arterial System

Arterial thrombosis is often secondary to atherosclerotic changes in
the vessel wall. The accumulation of atherosclerotic material leads
to a gradual narrowing of the vessel. When in this process the internal layer of the vessel wall is disrupted, the internal surface
looses its normally smooth appearance. On the resulting rough surface,
thrombosis can easily occur leading to a rapid progression of the
narrowing of the vessel or even to total occlusion.

The thrombus consists of fibrin, blood platelets and other elements
of the blood. Smaller or larger parts of the thrombus can be dislodged
and reach the brain as arterial emboli (see below). Other less frequent causes for thrombus formation are an abnormality of blood com-
position, inflammation of the vessel wall and traumata (e.g. per-
cutaneous angiography) of the vessel. The location of atherosclerotic
changes shows a decreasing frequency in the following order of
sequence:
(a) the bifurcation of the common carotid artery into the internal
    and external carotid arteries

(b) the confluence of left and right vertebral arteries

(c) the proximal part of the middle cerebral artery.

Much less frequently do such changes occur in the intracranial part of the internal carotid artery, the subclavian artery, the innominate artery, the common carotid arteries and the vertebral arteries (Fig. 1.10).

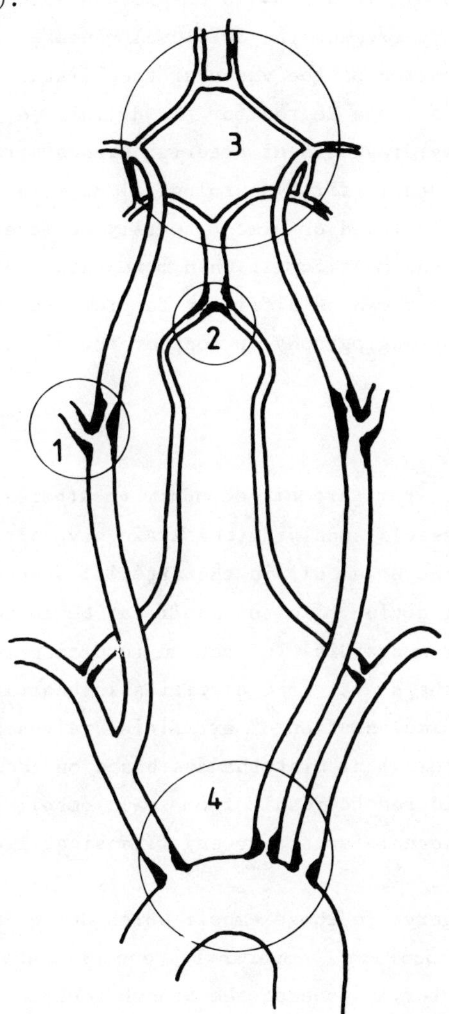

FIG. 1.10 Sites of predilection for atherosclerotic narrowing in the large arteries, in order of decreasing frequency.

Presumably a bruit perceptible at auscultation of the vessel, points to a stenosis of more than 50% of the original internal lumen. It will be obvious that bruits cannot occur when a vessel is totally occluded. Although a stenosis of 50% or more can produce a measurable pressure gradient over the stenosis it is a fact that in most cases a reduction in cerebral blood flow is only measurable when stenosis of the supplying artery exceeds the 80% level (Spencer and Reid, 1979). Probably vasodilatation of the vascular tree distal to the stenosis can compensate to a large degree for the diminished input pressure. This makes an early diagnosis of arterial stenosis rather difficult: most patients show their first neurological deficit only when a vessel is either totally occluded or almost totally occluded.

Stenoses are often multifocal. When more than one stenosis occur in the same vessel it can be difficult to prove the more distally located stenosis by angiography or Doppler techniques.

1.5.3 <u>Embolism</u>

Emboli originating from thrombi secondary to atherosclerotic changes in the arterial vessels supplying the brain have already been discussed. The main source of emboli in the arterial system of the brain is the heart. Thrombi giving rise to emboli can be formed in the cavities of the heart after myocardial infarction resulting in damage to the intimal layer of the heart, irregularities in heart rhythm, valvular disease and after implantation of artificial valves. At present most short-lasting neurological disturbances based on localized cerebral ischemia are considered to result from small emboli generated from lesions in the internal carotid artery (Transient Ischemic Attacks = TIA's).

A special category are those emboli which do not enter one of the brain vessels but occlude (temporarily) one of the vessels of the eye (the ophthalmic artery is one of the branches of the internal carotid artery). This causes a transient loss of vision of one eye (= amaurosis fugax). This is an important warning sign that emboli are circulating in the arterial system and should be appreciated as a very strong indication for exhaustive diagnostic procedures to evaluate the origin of these emboli. It is not possible to summarize the clinical effects of occlusion of cerebral vessels. To a large degree the clinical outcome is determined by the existing or developing collateral circulation.

1.5.4   Spasms

Severe vascular spasms may result in an insufficient blood supply of (part of) the brain. The benign forms of vascular spasm (e.g. as encountered in migraine) rarely lead to permanent neurological deficit but in isolated cases this is possible.

The most severe spasms, often resulting in severe impairment of cerebral circulation, are encountered in patients with subarachnoidal hemorrhage. This is a bleeding from an abnormal congenital or acquired outpouching of the arterial wall (aneurysm). Such bleeding may in itself have disastrous results or even cause sudden death. If the patient survives for a period severe cerebral arterial spasms may occur, probably induced by the blood lying around the cerebral arteries. These spasms are sometimes more important to the clinical outcome than the bleeding itself.

1.5.5   Arteriovenous Malformations

Arteriovenous malformations are a developmental disorder in which direct connections exist between the arterial and venous system without the normal system of small arteries, capillaries and small veins. These pathological connections can be quite large. Moreover, the vessel wall of these structures has often an abnormal structure prone to rupture and bleeding.

Apart from this the locally diminished cerebrovascular resistance gives rise to shunting of the blood from the arterial to the venous system without any participation of this blood in the normal metabolic process. This shunting of the blood may take such proportions that blood is withdrawn from neighbouring areas leading to ischemic disturbances in a smaller or larger part of the surrounding brain. It will be understandable that this kind of ischemic brain disease is the only one where blood flow velocity in the common carotid artery as measured by Doppler techniques is not diminished but normal or even enhanced.

This short survey of ischemic brain disease is certainly not complete. There are many other causes for primary cerebral ischemia (e.g. kinking and tortuosity of the large arteries) or secondary cerebral ischemia (intracerebral tumours, intracerebral hemorrhage etc.). It may be sufficient, however, to outline that the origin and mechanisms of cerebrovascular insufficiency are complex and often difficult to explore and diagnose.

In our view the Doppler technique is a very valuable tool in the diagnosis of cerebrovascular disease and will be even more so when more sophisticated Doppler techniques are used in clinical routine.

1.6   CONCLUSIONS

In conclusion it can be stated that the anatomy and pathology as well as the physiology and clinical physiology of the cerebral blood flow are well known. Until now this knowledge has not yielded important clinical results but it has made the clinician more aware of the importance of arterial P_{CO_2}, intracranial pressure and mean arterial blood pressure in his treatment of patients with cerebrovascular disorders. There is also, at present, little pharmacological treatment of patients with ischemic brain disease. Drugs with a longlasting vasodilatatory effect do not exist. Some drugs have an impressive effect on the cerebral circulation (e.g. papaverine) but are clinically worthless because the effect is of very short duration and the drug has to be applied intra-arterially. To prevent the formation of thrombi (and emboli) consisting of thrombocytes, anti-platelet agents may be useful. In some cases the use of anti-coagulants can be considered.

Reconstructive surgery should be considered in those patients with ischemic brain disease due to a local obstruction of one of the arteries. This has been possible only in the carotid arteries. Early detection of these operable stenoses is the most important and most rewarding task for Doppler flow velocity measurement techniques. The latest possibility in vascular surgery is the so called "bypass technique". A branch of the external carotid artery is connected with a branch of the internal carotid artery thus creating a normally non-existent anastomosis which "bypasses" an obstruction in the internal carotid artery or one of its primary branches. The connection is made through a burr-hole in the skull.

From the point of view of Doppler technology it might be worthwhile to note that this artificial anastomosis can be easily studied by ultrasound. This is a simple way to prove the patency of such an anastomosis.

We will not end this chapter without a perhaps superfluous warning. The Doppler technique can be of value in studying the physiology and pathophysiology of the cerebral circulation but it should be kept in mind that blood flow velocity rather than blood volume flow is measured with this technique. Blood flow velocity measurements, however, have been shown to be an appropriate tool in the detection of lesions in the extra-cranial arteries. On the other hand the development of an ultrasound device, having the possibility to measure accurately volume flow through a distant artery, would be a major breakthrough for studying intracranial disorders. At present cerebral blood flow can only be measured with the use of radioactive isotopes.

REFERENCES

Bayliss, W.M. (1902). On the local reactions of the arterial wall to changes of internal pressure. J. Physiol. 28, 229-231.

Fields, W.S., Bruetman, M.E. and Weibel, J. (1965). Collateral circulation of the brain. Th. Williams and Wilkins Comp. Baltimore.

Gabrielsen, T.O. and Greitz, T. (1970). Normal size of the internal carotid, middle cerebral and anterior cerebral arteries. Acta Radiol. (Diagn.) 10, 1-10.

Olesen, P.J. (1971). Contralateral focal increase of cerebral blood flow in man during arm work. Brain 94, 635-646.

Spencer, M.P. and Reid, J.M. (1979). Quantitation of carotid stenosis with continuous wave (CW) Doppler ultrasound. Stroke 10, 326-330.

ADDITIONAL LITERATURE RELATED TO THE SUBJECT

McHenry, Jr., L.C. (1978). Cerebral circulation and stroke. Warren
   H. Green, Incl. St. Louis, Mo.

Krayenbühl, H.A. and Yasargil, M.G. (1968). Cerebral angiography.
   Butterworths, London.

Lehrer, H.Z. (1969). Cervical internal carotid artery calibre:
   variation with arterial occlusions. Acta Neurol. Scand. 45, 277-291.

Lie, K.T.A. (1968). Congenital anomalies of the carotid arteries.
   Thesis, Excerpta Medica Foundation, Amsterdam.

Meyer, J.S. (1975). Modern concepts of cerebrovascular disease.
   (Edited by J.S. Meyer). Spectrum Publications, New York.

Töndury, G. (1959). Angewandte und topografische Anatomie. Georg
   Thieme Verlag, Stuttgart.

Weibel, J. and Fields, W.S. (1965). Tortuosity, coiling and kinking
   of the internal carotid artery. I. Etiology and radiographic anatomy. Neurology 15, 7-18.

CHAPTER 2

Some Fluid Dynamical Aspects of Arterial Flow

M. E. H. van Dongen, *and* A. A. van Steenhoven

2.1   INTRODUCTION

In this chapter we will describe some basic fluid-dynamical aspects
of velocity profiles as measured in the arterial system for example
with ultrasonic Doppler instruments. These instruments have a tremen-
dous potential because they can record almost instantaneously
velocity profiles without the necessity of inserting a sensing device
into the vessel. Such a velocity profile shows the distribution of
the velocity of fluid particles flowing through a cross-section of an
artery. As there are three velocity components, a single plot of the
distribution of one velocity component only then yields valuable
information if the flow behaviour may be regarded as unidirectional.

The recorded velocity within a vessel at one instant is in general
not uniform over a cross-section. The best known example of a velocity
profile is that of Poiseuille flow, where the velocity parabolically
ranges from zero at the wall to twice the cross-sectional average at
the center-line. It should be noted that Poiseuille flow only exists
in regions of steady flow in a straight pipe, far away from the entrance, in which no sources of flow disturbance, such as stenoses,
are present. These conditions are generally not met in the systemic
arteries and therefore a great variety of profiles is observed in
these vessels. In Fig. 2.1 examples of profiles are schematically
shown, together with the ratio of the maximum velocity to the average
velocity over the cross-section. They all differ significantly from
those associated with Poiseuille flow.

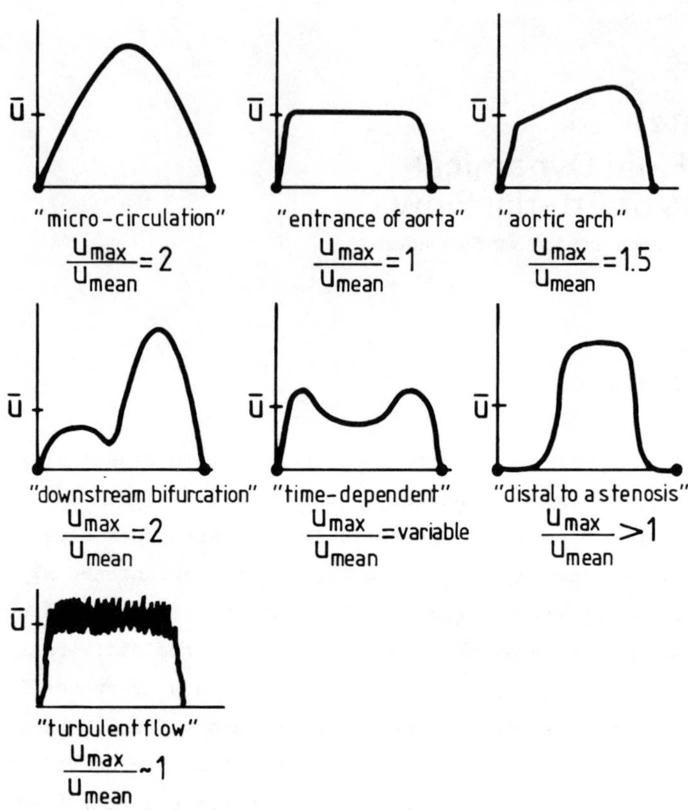

FIG. 2.1   Examples of velocity profiles as measured
in the arterial tree. Based upon figures of Caro
et al, 1979.

In this chapter we will discuss the circumstances which influence the
velocity profile, like inlet-phenomena, viscosity effects of blood,
flow unsteadiness, vessel tapering and curvature, bifurcations and
wall elasticity. Because of the scope of this book, we will mainly
confine ourselves to the cervical carotid arteries. The characteristic values of the hemodynamic parameters for this artery in humans
are summarized in Table 2.1. Special attention will be paid to the
changes in the velocity profile around a stenosis as defined in
Fig. 2.2, because of the increasing evidence that disturbances in
the velocity profile do occur at relatively slight degrees of artery

narrowing (see chapters 9 and 10). These disturbances will be discussed in relation to some preliminary results on the velocity profiles as recorded around a model stenosis, using a multi-channel
pulsed Doppler system as designed by Hoeks and co-investigators
(1981). For more detailed information on the subject matter we refer
to textbooks by Caro et al (1978), McDonald (1974) and Pedley (1979).
The first book especially discusses the fluid mechanical aspects
relevant to the present chapter in a thorough and multi-disciplinary
way.

The symbols used in this chapter are listed and defined at the
end after the references.

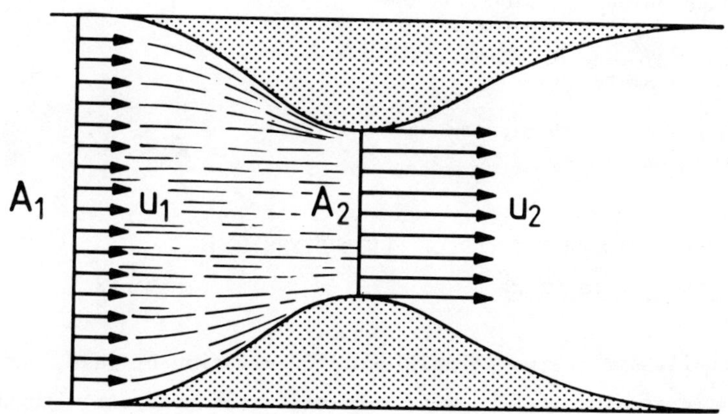

FIG. 2.2   Definition of a stenosis. The cross-sectional
areas are symbolized by A and the axial velocities by μ.
The degree of a stenosis is defined by $(1-A_2/A_1) \times 100\%$

TABLE 2.1 : Characteristic values for the human carotid artery

Variable	Value		Source
Angular frequency ω (first harmonic)	6	rad/sec	
Diameter d	8×10^{-3}	m	Learoyd (1966)
Wall thickness	4×10^{-4}	m	McDonald (1974)
Young's modulus	4.5×10^{5}	N/m^2	Learoyd (1966)
Pressure amplitude (first harmonic)	3×10^{3}	N/m^2	Arndt (1968)
Peak velocity	0.50	m/s	Mills (1970)
Velocity amplitude (first harmonic)	0.25	m/s	
Blood viscosity μ	3.8×10^{-3}	kg/m/s	McDonald (1974)
Blood density ρ	1.06×10^{3}	kg/m^3	
Peak Reynolds number $Re = \rho \bar{u} d / \mu$	1100		
Womersley's parameter $\alpha = \frac{1}{2} d \sqrt{\dfrac{\rho \omega}{\mu}}$ (first harmonic)	6		

2.2 SOME BASIC PRINCIPLES OF FLUID DYNAMICS

2.2.1 Conservation Of Mass

In order to illustrate the principle of conservation of mass, consider flow in an elastic tube with a cross-section varying with the
axial coordinate x (Fig. 2.3). We define two arbitrary cross-sections
perpendicular to the tube axis with areas A_1 and A_2, and denote the
average velocities in the axial direction by $\bar{u}_1$ and $\bar{u}_2$. The tube
volume bounded by the two cross-sections is V. The flow entering this
volume per unit time Q_1 equals $\bar{u}_1 A_1$ and a volume flow Q_2 is leaving
the volume at position 2. The principle of conservation of mass for
an incompressible fluid, which means that its mass density ρ is constant, states that the difference between Q_1 and Q_2 is equal to the

rate of change of the volume V:

$$\bar{u}_1 A_1 - \bar{u}_2 A_2 = \frac{dV}{dt} \qquad\qquad 2.1$$

For steady flow, the volume flow is obviously independent of the axial coordinate. Sometimes it is possible to neglect the dV/dt-term also for unsteady flow conditions, e.g. when the walls are relatively rigid and the cross-sectional area changes appreciable over a short distance such as in the case of a severe stenosis.

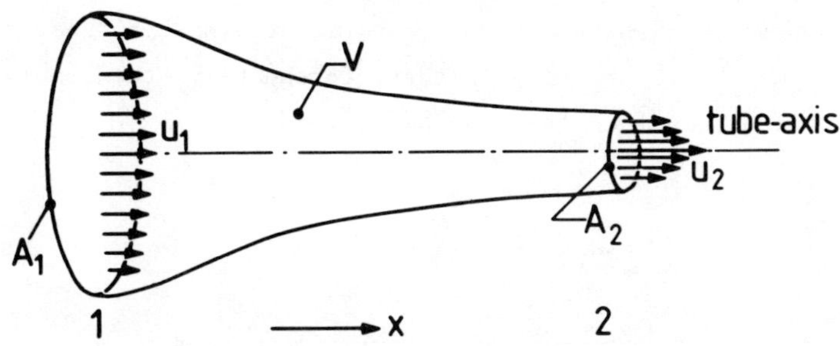

FIG. 2.3   Schematical drawing of an elastic tube with cross-sectional area A and axial velocity u. The tube volume bounded by the two cross-sections is V; x is the axial coordinate.

2.2.2   Forces Associated With Fluid Flow

All velocity profiles have one property in common, the fluid velocity at the wall is always zero. This is known as the so-called no-slip condition. Obviously, the wall exerts some force on the fluid particles adjacent to it. More generally fluid particles exert a shear force on neighbouring fluid particles moving in the same direction with a different velocity. This shearing force per unit area is called the shear stress. The magnitude of the shear appears to depend on the velocity gradient or shear rate $\partial u/\partial y$ where y denotes a space coordinate perpendicular to the flow direction.

If the shear stress σ is proportional to the shear rate,

$$\sigma = \mu \frac{\partial u}{\partial y} \qquad\qquad 2.2$$

the fluid is called Newtonian, and μ is called the viscosity of the fluid. Examples of Newtonian fluids are water and glycerine. Blood exhibits a more complex behaviour which can be attributed to the presence of the deforming and rotating red blood cells. At zero shear rate, the red blood cells form a bridge-like structure within the vessel, preventing blood flow until a certain critical stress, the yield stress, is exceeded. Blood viscosity decreases with increasing shear rate until an almost constant value is attained at shear rates typical for the larger arteries so that, fortunately, the complexity of non-Newtonian behaviour then can be avoided.

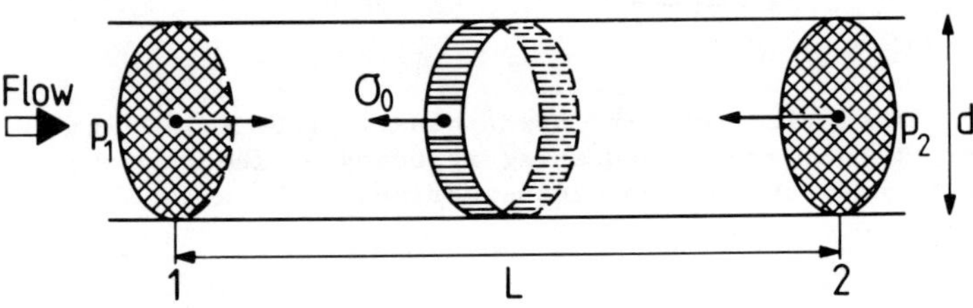

FIG. 2.4　　Forces acting on a tube element with diameter d and length L. The pressure is denoted with p and the shear stress with σ_0

Obviously, shear forces cannot be the only forces acting on the moving fluid and some balancing force has to be present whenever a state of equilibrium has to be maintained. In many situations this force is exerted by the pressure p, which is the force per unit area, acting on a given surface element in a direction perpendicular to it. This balance of forces is illustrated by considering flow in a straight tube element of diameter d and length L, remote from the entrance so that the velocity profile is not dependent on the axial coordinate (Fig. 2.4). The shear force exerted by the wall on the fluid in the negative x-direction has here to be counterbalanced by the net pressure force, pointing in cross-section 1 to the right and in cross-section 2 to the left. The balance of forces, therefore, becomes:

$$\tfrac{1}{4}(p_1-p_2)\ \pi d^2 = \sigma_o\ \pi d\ L \qquad\qquad 2.3$$

or

$$\frac{p_1 - p_2}{L} = \frac{4\sigma_o}{d} \qquad\qquad 2.4$$

The pressure difference per unit length, or the pressure gradient, appears to be the driving force for this particular situation. This type of flow is of course the well-known Poiseuille flow and the famous Poiseuille formula can be found by realizing that formula 2.2.3 is not only applicable to the whole tube element of radius $\tfrac{1}{2}d$, but to any circular cylinder with radius $r \leqslant \tfrac{1}{2}d$. It is found then that the velocity profile is parabolic:

$$u = 2\ \bar{u}\ \left[1-\left(\frac{r}{\tfrac{1}{2}d}\right)^2\right] \text{ and } \sigma_o = \left.-u\ \frac{\partial u}{\partial r}\right|_{r=\tfrac{1}{2}d} = 8\ \mu\ \frac{\bar{u}}{d} \qquad\qquad 2.5$$

Inserting this relation into equation 2.3 then yields the Poiseuille formula:

$$\frac{p_1 - p_2}{L} = 32\ \mu\ \frac{\bar{u}}{d^2} = 128\ \mu\ \frac{Q}{\pi d^4} \qquad\qquad 2.6$$

In the human body pressure can vary not only because of the pumping action of the heart, but also because of gravity. In a resting fluid, pressure varies with height according to the hydrostatic law:

$$p = -\rho g\ (z-z_o) \qquad\qquad 2.7$$

where g is the acceleration of gravity, z is the coordinate in the positive vertical direction and z_o is some reference height.

In a closed system, the reference height is a constant and the driving forces for the fluid motion can be found by interpreting the pressures p_1 and p_2 as the excess pressures with respect to the hydrostatic values given by equation 2.7.

We now return to the simple flow situation of Fig. 2.4 and wonder what will happen when we increase the pressure gradient. Then there will be a net force acting on the fluid, and consequently the fluid will accelerate. According to Newton's law equation 2.3 then has to be modified as:

$$\tfrac{1}{4}\rho\pi d^2 L\,\frac{d\bar{u}}{dt} = \tfrac{1}{4}(p_1-p_2)\pi d^2 - \sigma_o\,\pi dL \qquad\qquad 2.8$$

or

$$\rho\frac{d\bar{u}}{dt} = -\frac{(p_2-p_1)}{L} - \frac{4\sigma_o}{d} \qquad\qquad 2.9$$

This equation, however, cannot be solved in general due to the dependence of the left hand side on the unknown velocity profile. In such a case use has to be made of a more fundamental law, the so-called Navier-Stokes equation, which describes how a fluid element of infinitesimal size is accelerated due to the pressure gradient and the net viscous forces exerted by the neighbouring fluid particles. Its structure is similar to equation 2.9:

density x acceleration = - pressure gradient + net viscous force 2.10

It is common in fluid dynamics to refer to the acceleration term as the inertia force term. Two types of inertia forces can be distinguished, associated with two types of fluid acceleration. The first one we have already met, the local acceleration, when on a given place the local velocity is time dependent. But also in steady flow a fluid particle can accelerate along its path as happens for example in the converging part of a stenosis. Such a convective acceleration affects the pressure distribution, which is most easily demonstrated by Bernoulli's theorem.

2.2.3   <u>Bernoulli's Theorem</u>

We now consider an idealized incompressible flow, in which the shear
forces can be neglected. In general this restriction certainly does
not hold for the human circulatory system as a whole, but locally it
can be a good approximation. We further introduce the concept of a
streamline, which is defined so that at any point the velocity vector
is tangential to the streamline at a given instant as illustrated in
Fig. 2.5.

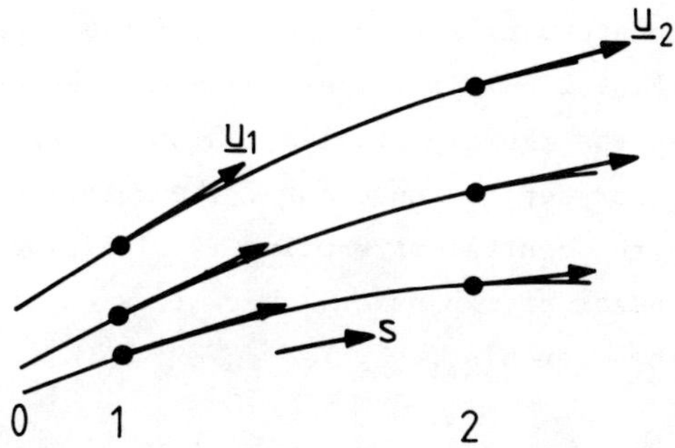

FIG. 2.5   Streamlines at a given instant.
At any point the velocity vector $\underline{u}$ is
tangential to the streamline. The coordinate along it is denoted with s.

From the Navier-Stokes equation, it is possible to derive that, as
long as viscous forces are negligible, the total pressure head,
defined as:

$$H = p + \tfrac{1}{2}\rho u^2 + \rho \int_{s_o}^{s} \frac{du}{dt}\, ds \qquad\qquad 2.11$$

is constant along a streamline, where s is a distance measured along
the streamline and s_o corresponds to a reference point. This is known
as Bernoulli's theorem. It implies for example that for steady flow
in a convergent-divergent tube, a velocity increase causes a pressure

decrease and vice-versa. It should be noted that the total pressure head may vary considerably from streamline to streamline.

The viscous forces tend to have a decreasing effect on the total pressure head in the downstream direction, which means that part of the mechanical energy will be converted into heat. As an example we will discuss schematically the steady flow through two types of stenoses as shown in Fig. 2.6. The shape of the stenosis appears to have an important effect on the loss in total pressure head along the central streamline. If the diverging part of the stenosis expands gradually (Fig. 2.6a), the loss in total pressure head may be relatively small and is not very different from the loss that would have been observed if no stenosis was present. When the expansion is abrupt, however, (Fig. 2.6b), the loss in total pressure head may be appreciable because the fluid particles can no longer follow the wall-contour, and flow separation occurs. The velocity and pressure distribution along the central streamline is also given in Fig. 2.6. In the case with an abrupt expansion, there is a considerable pressure drop across the stenosis.

2.3   FACTORS INFLUENCING THE VELOCITY PROFILE

2.3.1   The Formation Of A Boundary Layer

We now consider the steady flow in a long circular tube near the entrance. There the velocity profile is almost flat except in a very thin layer close to the walls, since the velocity at the wall has to be zero. This thin layer, where the velocity profile is directly affected by the viscous forces, is called a boundary layer. Downstream, this boundary layer becomes thicker as is illustrated in Fig. 2.7, while the velocity in the flat core of the tube is increasing. After some length, the entrance length, the boundary layer thickness approximately equals the tube radius and the velocity profile becomes that of fully developed Poiseuille flow. Since the shear force exerted by the wall is proportional to the shear rate, i.e. $\partial u/\partial r$ at the wall, this shear force is decreasing in the downstream direction until the constant value of Poiseuille flow has been attained. This is one of the reasons that the pressure drop per

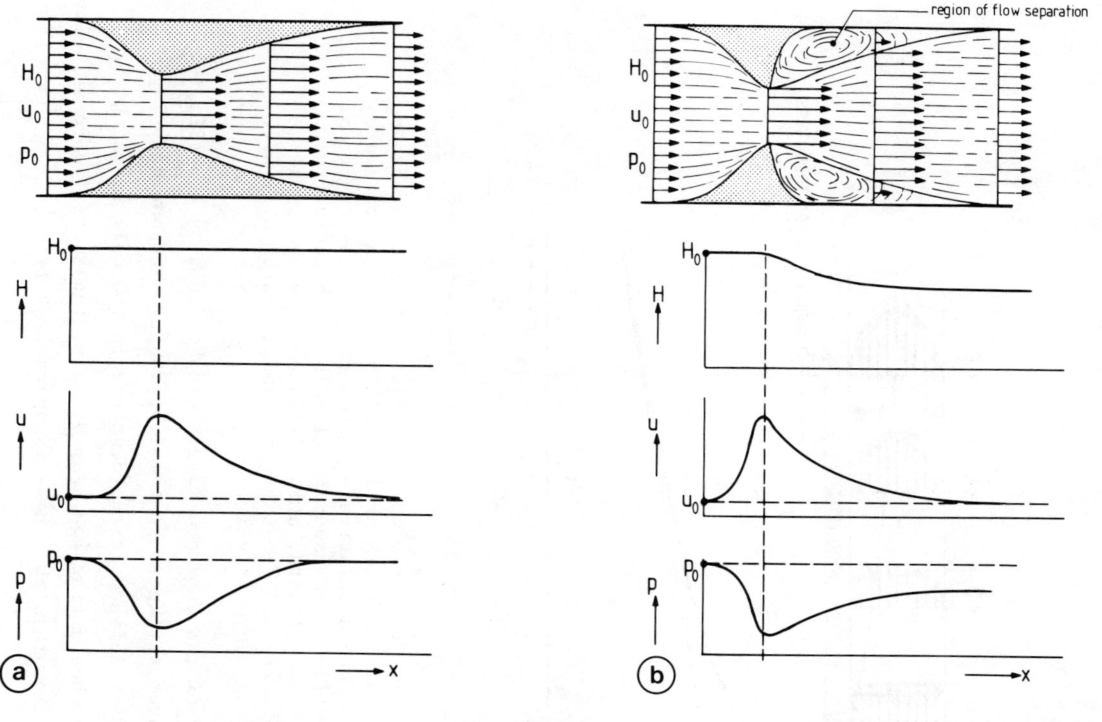

FIG. 2.6   a. Loss of total pressure head along the central streamline with a gradual expansion. The dotted lines are streamlines, while H, u, p and x symbolize the pressure head, velocity, pressure and axial distance, respectively.

b. Loss of total pressure head along the central streamline with an abrupt expansion. For symbols see Fig. 2.6a.

unit length, i.e. the pressure gradient, in the entrance region of the tube exceeds that of fully developed Poiseuille flow as is shown in Fig. 2.7.

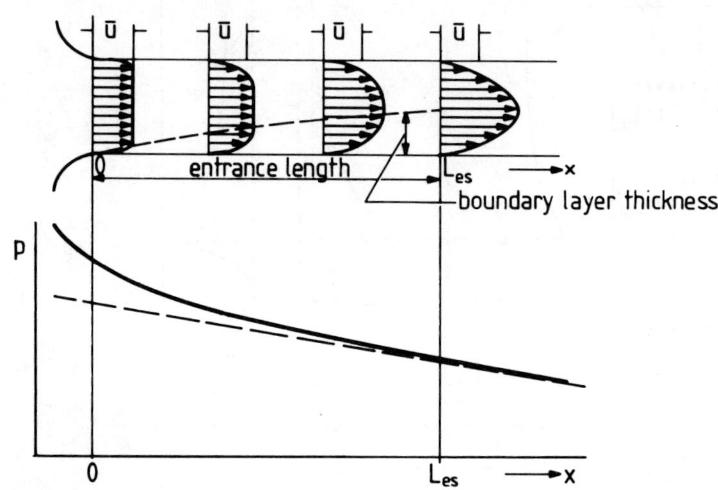

FIG. 2.7   The development of laminar flow through a circular tube. Starting with a blunt velocity profile and a relatively large rate of pressure drop. Finally the Poiseuille flow is established. The entrance length, axial distance, average axial velocity and pressure are symbolized by L_{es}, x, ū and p, respectively.

An estimate of the entrance length can be found by equating the boundary layer thickness to the tube radius. The boundary layer thickness δ is defined as the distance to the wall at which the velocity reaches a fraction, say 0.99, of the central core velocity. For steady flow this thickness is approximately:

$$\delta_s = k \sqrt{\frac{\mu L}{\rho \bar{u}}} \qquad\qquad 2.12$$

where k is a numerical constant, L the distance to the entrance plane and $\bar{u}$ the average velocity over a cross-section. Equating δ_s to the tube radius $\frac{1}{2}d$, we find for the entrance length L_{es}:

$$L_{es} = \frac{1}{4k^2} \, d \, \frac{\rho \bar{u} d}{\mu} \qquad\qquad 2.13$$

For the numerical factor $\frac{1}{4k^2}$ one usually takes a value of about 0.02. The term $\frac{\rho \bar{u} d}{\mu}$ is called the Reynolds number and appears to be of fundamental importance for the description of the flow. It should be mentioned that the formula 2.13 is valid for Reynolds numbers between 10 and 2500. Substituting the physiological values for the carotid artery as summarized in Table 2.1 one finds a Reynolds number of 1100 and an entrance length of 18 cm. This clearly indicates that entrance effects cannot be neglected when studying velocity profiles in the carotid artery.

A situation similar to that of the entrance region of a tube can be found downstream to any place where the velocity profile has been disturbed for some reason. Examples are bifurcations, bends and constrictions. Viscous action then tends to restore the Poiseuille profile downstream to such a disturbance. The actual situation in the human body is of course more complex because of the unsteady character of the flow.

2.3.2 <u>Reynolds Number, Turbulent Flow</u>

The Reynolds number, previously defined as $\rho \bar{u} d / \mu$, is a dimensionless quantity, which plays an important role in characterizing a flow situation. It can be interpreted as the ratio of the steady inertia force, $\rho \bar{u}^2$, and the shear force $\mu \bar{u}/d$. A low Reynolds number refers to a situation where the viscous forces are dominant, while a large Reynolds number indicates that the inertia terms are significant.

As was demonstrated by Reynolds, some 100 years ago, pipe flow at very high Reynolds numbers ($\gg$ 1000) differs completely from Poiseuille flow. The pressure gradient is no longer linearly related to the flow rate, but is approximately proportional to its square. The fluid motion far from the entrance of the tube is not any more

42

orderly and strictly steady, but exhibits fluctuations, which are
random in amplitude, frequency and direction. This type of flow is
called turbulent as distinct from the laminar flow for the lower
Reynolds numbers. The Reynolds number at which transition from laminar to turbulent flow occurs is somewhat dependent on the experimental circumstances and is in the order of 2000. The pressure of a
local constriction such as a stenosis may facilitate the laminar-
turbulent transition.

Although the velocity fluctuates at turbulent flow, it is possible
to define and to measure a time' average of the local velocity. The
velocity profile obtained in this way differs from Poiseuille flow
in the sense that the core is more flat and the velocity gradient at
the wall is steeper for turbulent flow. The peak velocity near the
axis is approximately 1.2 times the average velocity (Fig. 2.1). When
measuring this central streamline velocity in a time dependent flow
situation, for instance, by means of pulsed ultrasonic Doppler or hot
film anemometry, the turbulent eddies in the flow will be observed
as random fluctuations superimposed upon a mean flow velocity as is
shown in Fig. 2.8.

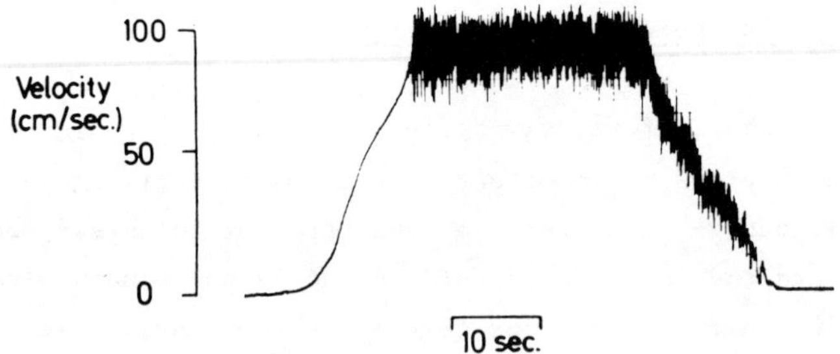

FIG. 2.8 Signal from hot-film needle probe in
pipe-flow. The flow was increased until turbulence occurred and then stopped. Peak Reynolds
number 9500. From: Nerem and Seed, 1972.

2.3.3   Unsteady Flow

Again we return to the long circular pipe far from the entrance and exit, in which now the pressure gradient is varied with time in a sinusoidal manner with angular frequency ω. When ω is very small, we observe a parabolic velocity profile at any instant throughout the cycle, typical for Poisseuille flow. In that case the unsteady inertia force of equation 2.9 can be neglected and at each site there is still an instantaneous balance between the pressure force and the shear force. When the frequency is high, the situation is different. In the core the viscous forces are relatively less important. At this site there is a balance between inertia force and pressure gradient, so that velocity and pressure gradient are 90 degrees out of phase. Close to the wall there is still a balance between pressure and viscous forces.

Similar to the situation near the entrance of a tube in steady flow, we can distinguish between an inviscid core and a thin layer close to the wall where viscous forces are important. The thickness of this unsteady boundary layer δ_t, depends on the frequency, the viscosity and the density of the fluid and is equal to:

$$\delta_t = k^{\mathbf{x}} \sqrt{\frac{\mu}{\rho\omega}} \qquad\qquad 2.14$$

$k^{\mathbf{x}}$ being a numerical factor, which is usually taken to be 6.5. We now can state more precisely that the flow shows an unsteady Poiseuille character when $\delta_t \gg \tfrac{1}{2}d$ and that a large inviscid core will be present if the ratio $2\delta_t/d \ll 1$. The inverse of this ratio, apart from the numerical factor, is usually called Womersley's parameter α, being defined as:

$$\alpha = \tfrac{1}{2}d \sqrt{\frac{\omega\rho}{\mu}} = k^{\mathbf{x}} \frac{d}{2\delta_t} \qquad\qquad 2.15$$

For the carotid artery (Table 2.1) α equals 6, so that viscosity affects the whole flow field. This is illustrated in Fig. 2.9, where the flow pattern is shown at distinct time intervals for $\alpha = 6$; the

44

mathematical treatment can be found for example in Womersley (1955).
The actual flow pattern in the carotid artery is different from that
of Fig. 2.9, since the steady component and the higher harmonics
were not taken into account.

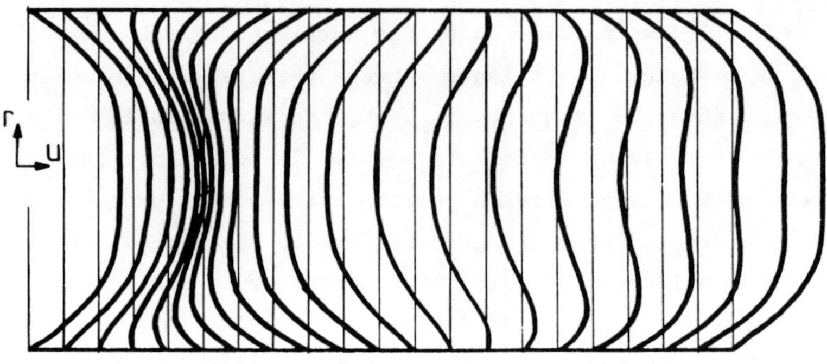

FIG. 2.9 The calculated radial (r) distribution of the axial
velocity (u) during a sinusoidal variation of the pressure
gradient for a rigid tube. The parameter values were those
of the carotid artery (see Table 2.1)

In the previous section we have given an estimate of the entrance
length for steady flow. The analysis of the entrance phenomena in
unsteady flow with a steady component is rather complicated since a
simple superposition of steady and unsteady flow fields is in general
not possible due to the non-linear character of the steady inertia
force term. Nevertheless, a rough estimate can be obtained by setting
the steady boundary layer thickness, evaluated at peak average velocity, equal to the unsteady boundary layer thickness, provided that
the latter is less than the tube radius. Combining equations 2.12
and 2.14 we then find:

$$L_{et} \simeq 3.4 \ \bar{u}/\omega \qquad\qquad 2.16$$

The peak velocity in the carotid artery is about 0.5 m/s (Table 2.1),
so that L_{et} is approximately 26 cm which is in the same order as the

entrance length for steady flow.

The unsteadiness of the flow does also affect the critical Reynolds number for the transition from laminar to turbulent flow. As the Womersley parameter α increases, the time available for the flow disturbances to develop decreases and for this reason the critical Reynolds number increases. This has been confirmed by experiments as is shown in Fig. 2.10 where the line Re = 250 α approximately separates laminar and turbulent flow conditions.

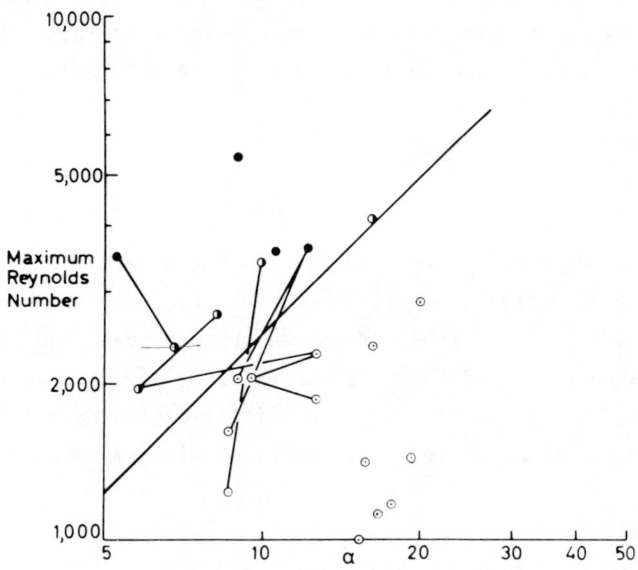

FIG. 2.10   Relation between the disturbance of flow and the peak Reynolds number and the Womersley parameter α for the descending thoracic aorta of dogs. The solid line corresponds to the relation Re = 250 α, ⊙ = undisturbed flow, ◑ = disturbed flow, ● = highly disturbed flow. From: Nerem and Seed, 1972.

Until now, we have assumed the walls to be perfectly rigid, so that the average velocity in a straight tube does not depend on the axial coordinate. In fact, the tube walls are elastic, resulting in a distension of the wall due to pressure variations. When the pressure

is time-dependent, this implies that wave phenomena occur. Pressure
and flow profiles are not steady, but propagate in the axial direction, and partly reflect from any geometrical disturbance. The speed
of propagation depends on the distensibility of the walls, i.e.
$\frac{1}{A}\frac{dA}{dp}$, and is about 4 m/s for the large arteries, the fluid velocity
itself being less than 1 m/s. The wavelength is in the order of 4 m,
so that the ratio of tube diameter and wavelength is much smaller
than unity. The velocity profile in a given cross-section in a long
circular elastic tube with properties typical for the arterial system
appears to be rather similar to the profile in a rigid tube, the
most striking difference being that the radial coordinate is strained
in proportion to the time varying value of the tube radius.

2.3.4 Geometrical Factors
2.3.4.1 Introduction

A complete description of the flow in the human vascular system is
difficult to give: the geometry is complex, the arteries have a
tapering cross-section, they are curved and bifurcate and above all,
the flow is unsteady and therefore affected by wave propagation and
reflections. In this section we will briefly discuss the influence
of local constrictions, stenoses, bends and bifurcations on the flow
field.

2.3.4.2 Constriction And Stenosis

A locally narrowed segment of an artery is called an arterial stenosis. Such a stenosis may lead to undesirable flow disturbances and
to serious circulation disorders. Stenoses frequently develop at
sites near the origin of branches and thus the geometry preceeding
the stenosis may be complex. For simplicity, we neglect the possible
influence of wall elasticity, which is a good approximation for a
severe stenosis, and will assume that the stenosis is situated in
between two straight tube segments.

The typical flow pattern that will be observed at a rather short,
abrupt and axi-symmetric stenosis has already been shown in Fig. 2.6b.

In the divergent part of the stenosis the streamlines diverge so
that the fluid particles decelerate, and hence an adverse pressure
gradient has to be present. Close to the wall the fluid particles
are not only retarded by the pressure gradient, but also by viscous
forces. Then flow separation occurs which means that the streamlines
near the wall cannot follow the wall contour and a region of closed
streamlines is present, the axial extent of which depends on the
Reynolds number. This is shown in Fig. 2.11, where the results of
numerical calculations by Wille (1979) are depicted for laminar flow.
It should be noted that for turbulent flow the tendency of the
streamlines to separate is less pronounced than for laminar flow,
although in turbulent flow separation certainly may occur.

Flow phenomena through and around stenoses have often been subject
to investigation. Recently a comprehensive review was given by Young
(1979). Steady flow phenomena around a model stenosis were studied
experimentally by Young and Tsai (1973a), Clark (1976a) and Tobin and
Chang (1976), and numerically by Deshpande et al (1976) and Wille
(1979). Geometrical effects of a stenosis were experimentally analyzed by Seeley and Young (1976) and Talukder et al (1977). Unsteady
flow experiments around a model stenosis were performed by Young and
Tsai (1973b), Clark (1976b), Cox et al (1979) and Newman et al
(1979), while Daley (1976) and Wille and Walløe (1981) performed
numerical studies on this subject. Model studies concerning the disorder of the flow distal to the stenosis and turbulence measurements
were performed by Clark (1976c, 1977), Khalifa and Giddens (1978),
Cassanova and Giddens (1978), Yongchareon and Young (1979) and
Deshpande and Giddens (1980).

We conclude this section on flow phenomena around a stenosis by
showing some preliminary results of oscillatory velocity profiles
measured a short distance from a slight model stenosis, using the
multi-channel pulsed Doppler system as designed by Hoeks et al (1981).
As shown in Fig. 2.12a, the experimental set-up consists of a long
flexible tube, through which a sinusoidal flow without DC component
is maintained. Since the flow direction reverses during the cycle
the velocity profile can be recorded upstream and downstream of the
stenosis without varying the site of measurement. The corresponding

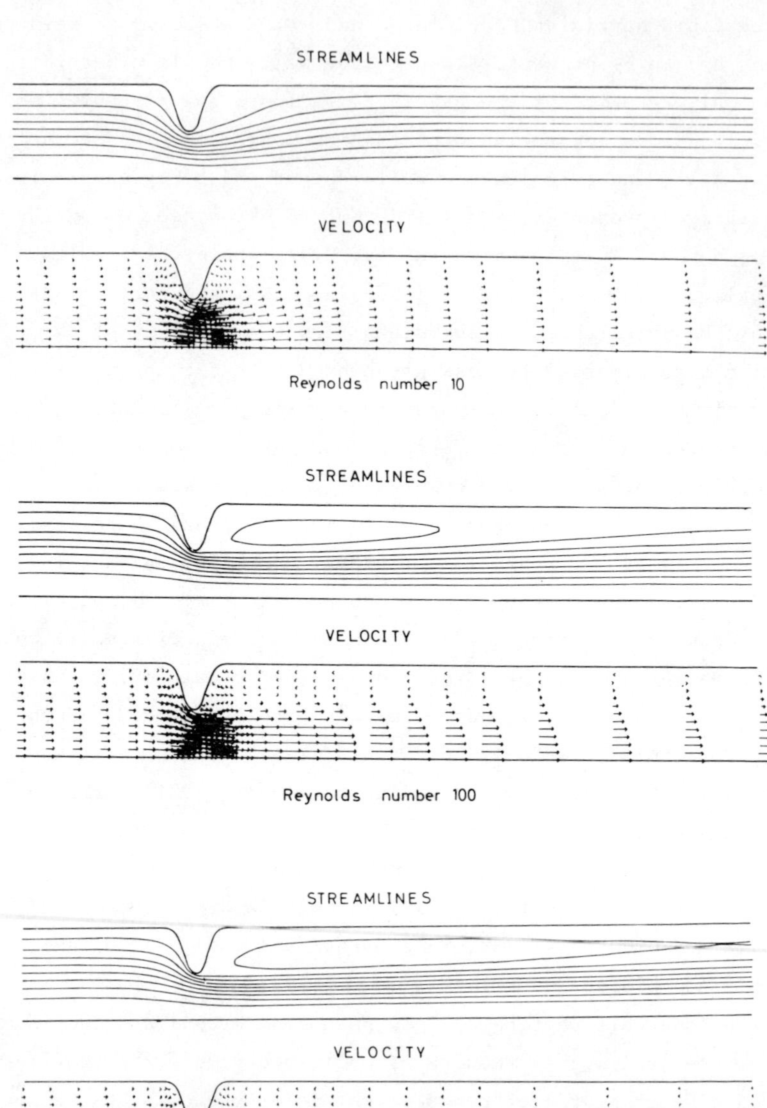

FIG. 2.11 Streamlines and velocity plots distal to a 75% stenosis, as found by finite-element calculations. From: Wille, 1979.

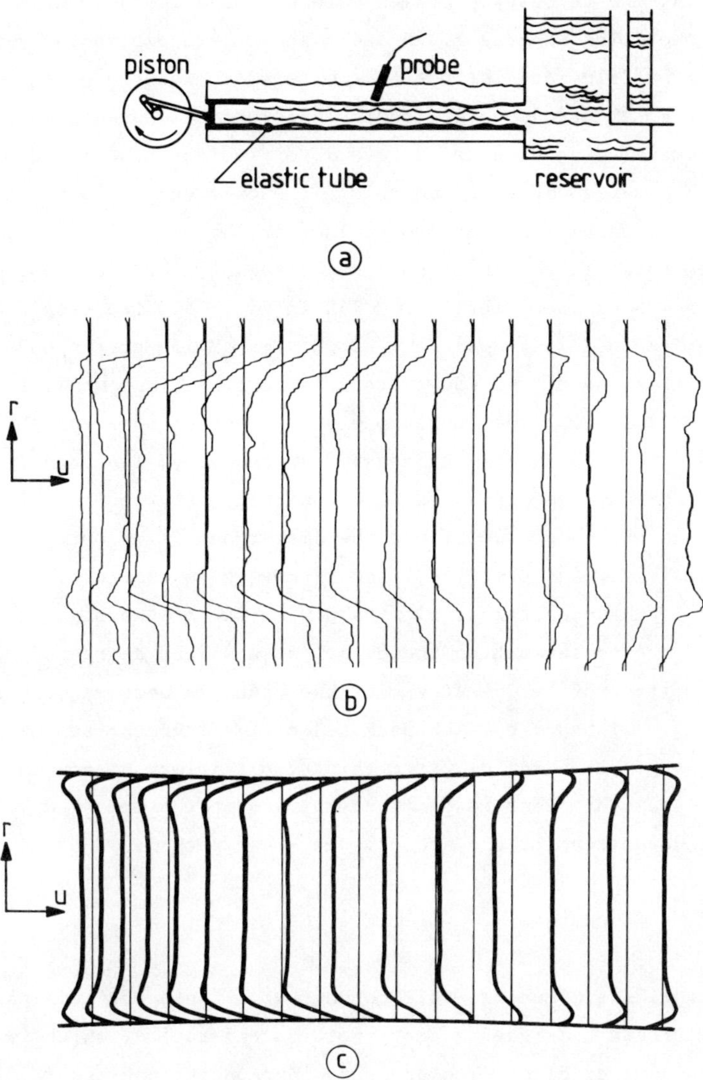

FIG. 2.12 A simple set-up for sinusoidal flow through
a straight flexible tube (a). Experimentally recorded velocity profiles (b) as measured with the
pulsed Doppler equipment of Hoeks et al, 1981 are
compared with those predicted from laminar theory
(c). (Re peak = 3500, α = 14 and peak velocity =
0.25 m/s).

peak Reynolds number is about 3500 while the Womersley parameter α
equals 14. The velocity profiles were measured in the central part
of the tube. The figures 2.12b and c show the experimentally recorded
velocity profiles at distinct time intervals, for the case without
a stenosis, simultaneously with the theoretically predicted profiles
using Womersley's calculation method for laminar flow (Womersley,
1957). The agreement is reasonable. It should be mentioned that it is
doubtful to assume the flow to be laminar, because of the high peak
Reynolds value, although it does not exceed the critical value predicted by 250 α (see section 2.3.1). In Fig. 2.13 the results to the
right of the model stenosis are shown; Fig. 2.13a corresponds to a
local stiffening of the tube without a constriction, while the
figures 2.13b and 2.13c correspond to a stenosis of 9 and 18%
(decrease in cross-sectional area), respectively. The frequencies
above 30 Hz have been reduced electronically, so that the turbulence
variations are suppressed. Note the difference in velocity profile
upstream (righthand profiles, flow is moving to the left) and downstream (lefthand profiles, flow is now moving to the right) of the
stenosis due to separation phenomena. At a slight degree of narrowing
the velocity profile downstream of the stenosis becomes more rounded
off (Figs. 2.13b and c). The peak velocities increase at this site
(cf Figs. 2.13a, b and c). From this result it may be concluded that
even at a slight stenosis distinct changes in the measured velocity
profile downstream to the stenosis may be expected.

2.3.4.3   Curvature

Most arteries are curved. This means that a fluid particle passing
such an artery undergoes a centripetal acceleration, which is caused
by a pressure gradient. The pressure increases along the radius of
curvature of the artery in the outward direction. As a result, the
axial velocity profile is distorted. Furthermore, due to viscous
effects a so-called secondary flow is generated: a vortex-like structure is superimposed on the flow in the axial direction. A typical
result of distortion of the profile in the aortic arch as measured
by Peronneau et al (1974) using a pulsed Doppler apparatus is shown

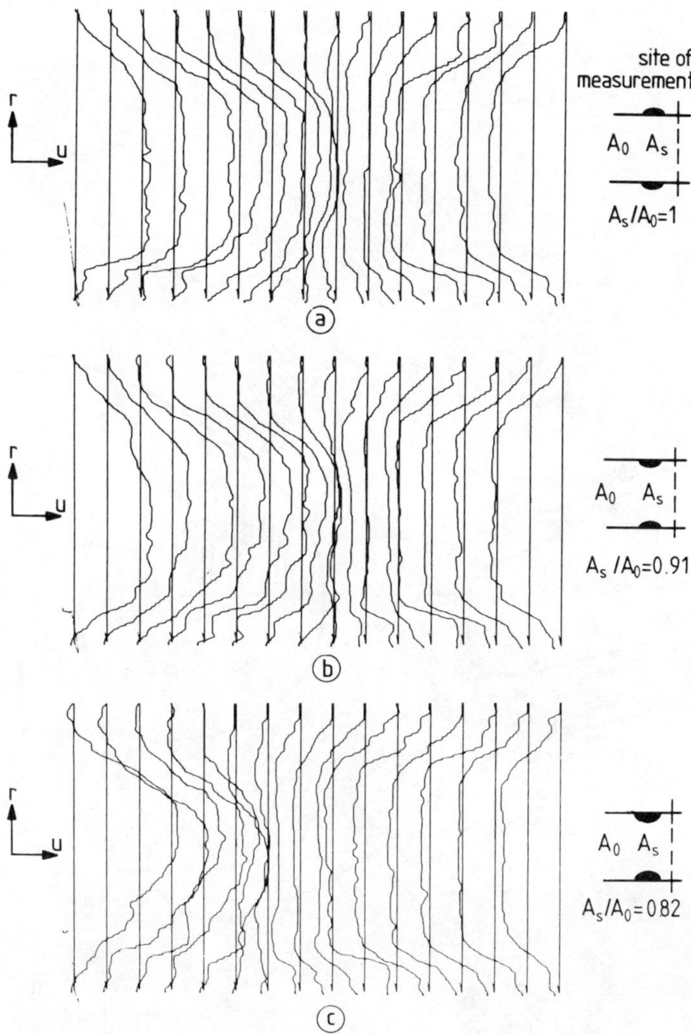

FIG. 2.13 The effect of stenosis on the velocity profile for various values of the cross-sectional area reduction A_s/A_0

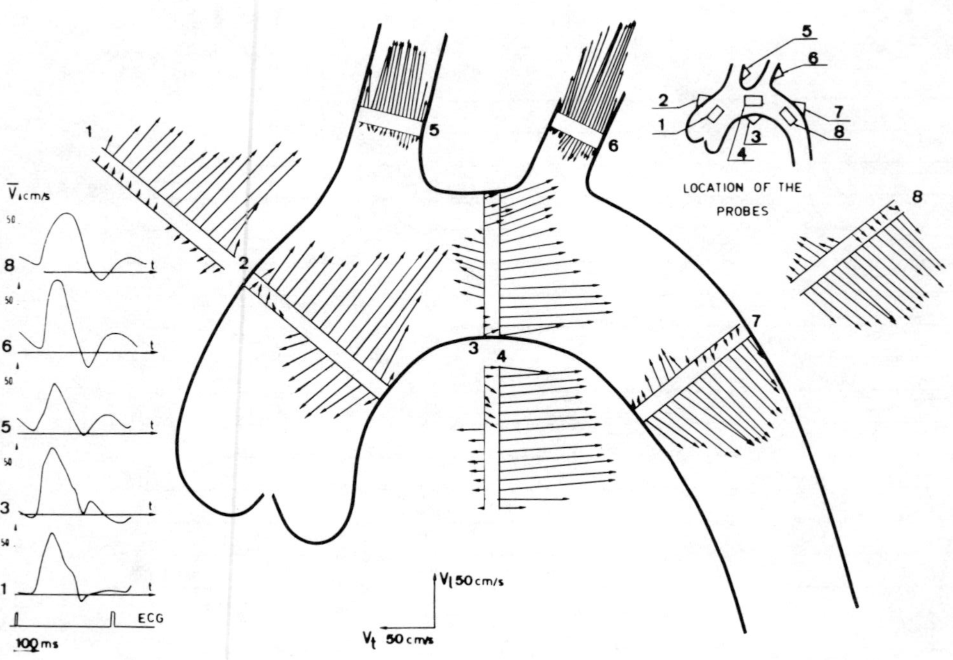

FIG. 2.14 Distribution of velocity vectors along the aortic arch, for peak forward and reversed flows. From: Peronneau et al, 1974.

in Fig. 2.14. Inlet phenomena play an important role in this situation.

Experimental model studies of this type of flow have been performed by Peronneau et al (1974) and Chandrau et al (1979b). Numerical analyses on this subject matter were done by Singh et al (1978) and Chandrau et al (1979a).

2.3.4.4   Bifurcation

Stenotic placques most often occur in or around bifurcations (Noon, 1977) and hence the influence of it needs to be accounted for in evaluating velocity profiles distal to a stenosis. The main effect of a bifurcation is that as the mainstream is divided into two parts, a boundary layer has to be developed at the inner walls of the bifurcating arteries and that consequently inlet phenomena play a dominant role. Furthermore, after a bifurcation separation of flow occurs, while also secondary flow phenomena are present (Caro et al, 1978).

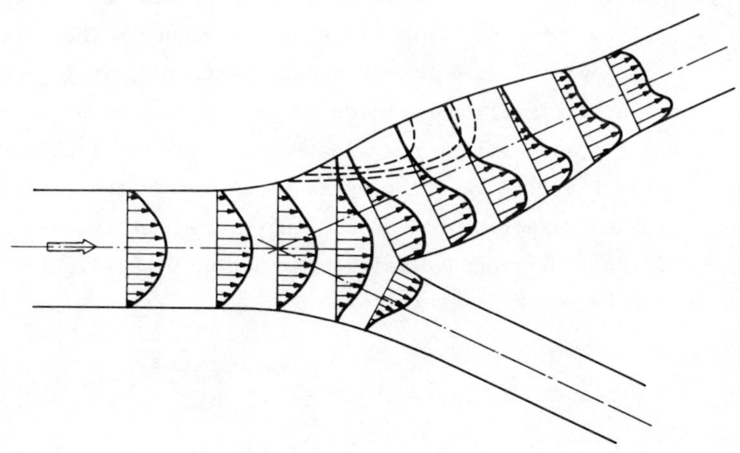

FIG. 2.15 Velocity profiles in the plane of bifurcation of a model carotid artery when Re = 400 (Dotted lines denote boundary of reversed flow zones in the plane of bifurcation. From: Balasubramanian, 1980.

54

The first two effects are schematically shown in Fig. 2.15, which is based upon laser-Doppler velocity measurements by Balasubramanian et al (1980).

Experimental work on flow phenomena around a model bifurcation has been performed by Ferguson and Roach (1972), Round et al (1977) and Balasubramanian et al (1980). Numerical work in this field has been done by Kandarpa and Davids (1976), Fernandez et al (1976), O'Brien and Ehrlich (1977) and Wille (1979).

2.4   CONCLUSIONS

From the foregoing it will be obvious that it is difficult to predict the velocity profile in a certain artery. Complicating factors have been shown to be the inlet phenomena, the unsteadiness of flow and the possible transition from laminar to turbulent flow. Furthermore, the complicated geometrical structure of the cardiovascular system like curvature and bifurcation of arteries and, in the case of athe-rosclerosis with the presence of stenoses, may cause flow separation and/or secondary flow phenomena, which disturb the axial velocity profile. Therefore, a lot of experimental and numerical work has been and will be performed to elucidate all those influences. The contri-bution of multi-channel pulsed Doppler instruments to this field is of great importance. Firstly because no sensing device needs to be inserted into the vessel which otherwise may disturb the flow and secondly because it will give a lot of in vivo information feeding future numerical and experimental model studies such as those focus-sed on the clinically important lesions without or with slight nar-rowing of the artery (see Chapter 10).

REFERENCES

Arndt, J.O., Klaushe, J., and Mersch, F. (1968). The diameter of the intact carotid artery in man and its change with pulse pressure. Pflügers Arch. 301, 230-240.

Balasubramanian, K., Giddens, D.P., and Mabon, R.F. (1980). Steady flow at the carotid bifurcation. In Biofluid mechanics. (Edited by D.J. Schneck). pp. 475-496, Plenum Press, New York-London.

Caro, C.G., Pedley, T.J., Schroter, R.C., and Seed, W.A. (1978). The mechanics of the circulation. Oxford University Press, Oxford-New York-Toronto.

Cassanova, R.A. and Giddens, D.P. (1978). Disorder distal to modeled stenoses in steady and pulsatile flow. J. Biomech. 11, 441-453.

Chandrau, K.B., Hosey, R.R., Ghista, D.N., and Vayo, V.W. (1979a). Analysis of fully developed unsteady viscous flow in a curved elastic tube model to provide fluid mechanical data for some circulatory pathophysiological situations and assist devices. J. Biomech. Eng. 101, 114-122.

Chandrau, K.B., Yearwood, T.L., and Wieting, D.W. (1979b). An experimental study of pulsatile flow in a curved tube. J. Biomech. 12, 793-805.

Clark, C. (1976a). The fluid mechanics of aortic stenosis-I. Theory and steady flow experiments. J. Biomech. 9, 521-528.

Clark, C. (1976b). The fluid mechanics of aortic stenosis-II. Unsteady flow experiments. J. Biomech. 9, 567-573.

Clark, C. (1976c). Turbulent velocity measurements in a model of aortic stenosis. J. Biomech. 9, 677-687.

Clark, C. (1977). Turbulent wall pressure measurements in a model of aortic stenosis. J. Biomech. 10, 461-472.

Cox, J.T., Van Hoften, J.D.A., and Hwang, N.H.C. (1979). Investigation of a pulsatile flowfield downstream from a model stenosis. J. Biomech. Eng. 101, 141-150.

Daly, B.J. (1976). A numerical study of pulsatile flow through stenosed canine femoral arteries. J. Biomech. 9, 465-475.

Deshpande, M.D. and Giddens, D.P. (1980). Turbulence measurements in a constricted tube. J. Fluid Mech. 97, 65-89.

Deshpande, M.D., Giddens, D.P., and Mabon, R.F. (1976). Steady laminar flow through modelled vascular stenoses. J. Biomech. 9, 165-174.

Ferguson, G.G. and Roach, M.R. (1972). Flow conditions at bifurcations as determined in glass models with reference to the focal distribution of vascular lesions. In Cardiovascular fluid dynamics II. (Edited by D.H. Bergel). pp. 141-156, Academic Press, London-New York.

Fernandez, R.C., De Witt, C.J., and Botwin, M.R. (1976). Pulsatile flow through a bifurcation with applications to arterial disease. J. Biomech. 9, 575-580.

56

Hoeks, A.P.G., Reneman, R.S., and Peronneau, P.A. (1981). A multi-
gate pulsed Doppler system with serial data processing. IEEE Transactions on Sonics and Ultrasonics, SU-28, 242-247.

Kandarpa, K. and Davids, N. (1976). Analysis of the fluid dynamic
effects on atherogenesis at branching sites. J. Biomech. 9, 735-741.

Khalifa, A.M.A. and Giddens, D.P. (1978). Analysis of disorder in
pulsatile flows with application to poststenotic blood velocity
measurement in dogs. J. Biomech. 11, 129-141.

Learoyd, T. (1966). Alterations with age in the viscoelastic properties of the human arterial walls. Circ. Res. 18, 278-292.

McDonald, D.A. (1974). Blood flow in arteries. Edward Arnold, London.

Mills, C.J., Gabe, I.T., Gault, J.H., Mason, D.T., Ross, J., Braun-
wald, E., and Shillingford, J.P. (1970). Pressure flow relation-
ships and vascular impedance in man. Cardiovasc. Res. 4, 405-417.

Nerem, R.M. and Seed, W.A. (1972). An in vivo study of aortic flow
disturbances. Cardiovasc. Res. 6, 1-14.

Newman, D.L., Westerhof, H., and Sipkema, P. (1979). Modelling of
aortic stenosis. J. Biomech. 12, 229-235.

Noon, G.P. (1977). Flow-related problems in cardiovascular surgery.
In Cardiovascular flow dynamics and measurements. (Edited by
N.H.C. Hwang and N.A. Normann). pp. 245-276, University Park Press,
Baltimore.

O'Brien, V. and Ehrlich, L.W. (1977). Simulation of unsteady flow
at renal branches. J. Biomech. 10, 623-631.

Pedley, T.J. (1980). The fluid mechanics of large blood vessels.
Cambridge University Press, Cambridge.

Peronneau, P.A., Hinglais, J.R., Xhaard, M., Delouche, P., and
Fhilippo, J. (1974). The effects of curvature and stenosis on
pulsatile flow in vivo and in vitro. In Cardiovascular applications
of ultrasound. (Edited by R.S. Reneman). pp. 203-215, North Holland/
American Elsevier, Amsterdam-London-New York.

Round, G.F., Pal, T.G., and Feuerstein, I.A. (1977). Viscous energy
dissipation for steady flow in models of arterial bifurcations.
J. Biomech. 10, 725-734.

Seeley, B.D. and Young, D.F. (1976). Effect of geometry on pressure
losses across models of arterial stenoses. J. Biomech. 9, 439-448.

Singh, M.P., Sinha, P.C., and Aggarwal, M. (1978). Flow in the
entrance of the aorta. J. Fluid Mech. 87, 97-120.

Talukder, N., Karayannacos, P.E., Nerem, R.M., and Vasko, J.S. (1977). An experimental study of the fluid dynamics of multiple non-critical stenoses. J. Biomech. Eng. 99, 74-82.

Tobin, R.J. and Chang, I.D. (1976). Wall pressure spectra scaling downstream of stenoses in steady tube flow. J. Biochem. 9, 633-640.

Wille, S.Ø. (1979). Numerical models of arterial blood flow. Ph.D. Thesis. Institute of Informatics, University of Oslo, Oslo.

Wille, S.Ø. and Walløe, L. (1981). Pulsatile pressure and flow in arterial stenoses simulated in a mathematical model. J. Biomed. Eng. 3, 17-24.

Womersley, J.R. (1955). Method for the calculation of velocity, rate of flow and viscous drag in arteries when the pressure gradient is known. J. Physiol. 127, 553-563.

Womersley, J.R. (1957). Oscillatery flow in arteries: the constrained elastic tube as a model of arterial flow and pulse transmission. Phys. Med. Biol. 2, 178-187.

Yongchareon, W. and Young, D.F. (1979). Initiation of turbulence in models of arterial stenoses. J. Biomech. 12, 185-196.

Young, D.F. (1979). Fluid mechanics of arterial stenoses. J. Biomech. Eng. 101, 157-175.

Young, D.F. and Tsai, F.Y. (1973a). Flow characteristics in models of arterial stenoses-I. Steady flow. J. Biomech. 6, 395-410.

Young, D.F. and Tsai, F.Y. (1973b). Flow characteristics in models of arterial stenoses-II. Unsteady flow. J. Biomech. 6, 547-559.

ACKNOWLEDGEMENTS

We gratefully acknowledge the numerical and experimental contribution of F.M. van de Vosse. We are indebted to A.J. Manders and H.G. Sonnemans for their help in preparing the illustrations.

LIST OF SYMBOLS

A, A_1, A_2	: cross-sectional area
d	: tube diameter
g	: acceleration of gravity
H	: total pressure head
k, k^x	: numerical constant
L	: length of tube segment
L_{es}, L_{et}	: entrance length for steady and time-dependent flow, respectively
Q_1, Q_2	: volume flow
P, P_0, P_1, P_2	: pressure
Δp	: pressure difference
r	: radial coordinate
Re	: Reynolds number: $Re = \sqrt{\dfrac{\rho u d}{\mu}}$
s, s_0	: coordinate along and reference point upon the streamline, respectively
t	: time
u, u_0	: axial velocity
$\underline{u}_1 / \underline{u}_2$	: velocity vector
$\bar{u}$, $\bar{u}_1$, $\bar{u}_2$	: axial velocity averaged over a cross-section
V	: volume
x, y, z	: Carthesian coordinates
α	: Womersley parameter : $\alpha = \tfrac{1}{2} d \sqrt{\dfrac{\omega \rho}{\mu}}$
δ_s, δ_t	: boundary layer thickness for steady and time-dependent flow respectively
ρ	: fluid density
σ, σ_0	: shear stress
ω	: angular frequency
μ	: fluid viscosity

CHAPTER 3

Some Hemodynamical Aspects of Large Arteries

M. P. Spencer *and* T. Arts

3.1   INTRODUCTION

This chapter deals with application of some concepts of hemodynamics to the relation between pressure and volume flow in a vessel segment. The latter relation is governed by fluid-dynamical laws, the most relevant principles of which are described in chapter 2. Unfortunately, a precise fluid-dynamic description of even simple flow phenomena is generally complicated. For the sake of clinical applicability, but at the cost of some accuracy in this chapter, the pressure flow relation in a vessel segment is modeled by a composite of discrete elements, each representing a separate aspect of fluid-dynamic behavior.

The arterial system is a many-branched elastic conduit for distribution of blood from the heart to all body tissues. Its calibre ranges from centimeters for the human aorta to micrometers (Henquell et al, 1976) for the capillaries. Over this wide range, the dynamics of each vascular segment may be described by various combinations of 3 kinds of elements (Spencer and Denison, 1963): resistance, inertance and compliance, collectively referred to as impedance. In relation to the description of a stenosis a fourth element, a non-linear stenotic resistance element, is introduced.

The symbols used in this chapter are listed and defined at the end following the references.

3.2   DISCRETE ARTERIAL SEGMENTS

The pressure-volume flow relation of discrete arterial segments is
described. Because of the common use of ultrasonic Doppler instrumentation the relation between pressure (p) and blood flow velocity
(v) is discussed as well.

3.2.1 Resistance

Linear resistance arises from viscous losses in the blood flowing
through the vessel segment. Hemodynamic resistance is analogeous
to electrical resistance, which we symbolize as—[____]—. Just as
electricity is impelled by a voltage difference, blood is driven by
a pressure difference between two points (Δp).

   Resistance (R) is defined as

$$R = \frac{\Delta p}{q}$$

    3.1

$$R = \frac{\Delta p}{v \cdot A_v}$$

    3.2

For a Newtonian fluid flowing through a cylindrical tube with a parabolic velocity profile this resistance equals

$$R = \frac{128 \, \eta \, l}{\pi \, d^4}$$

    3.3

The resistance of small blood vessels like small arteries and larger
arterioles is generally reasonably well described by equation 3.3.
However, the smaller the vessel, the less Newtonion the behavior
of blood is, and corrections for the viscosity have to be introduced. Equation 3.3 cannot be applied in case of a non-parabolic
velocity profile or a change in vessel diameter along the flow directions as is in stenoses and bifurcations in large arteries.

3.2.2   Pressure Drop Across A Stenosis

The pressure drop across a stenosis with steady flow is predominantly
determined by a viscous term Δp_v and a Bernoulli term Δp_k. The latter
term is related to acceleration of blood within the stenosis and turbulence distal to the stenosis. Dynamic inertial effects are considered to be of minor importance in this situation (see Section 3.2.
3.1).

In large blood vessels the viscous behavior of blood is considered
to be Newtonian, i.e. the ratio of shear stress and shear rate is
constant and equal to the viscosity (see equation 2.2). In the blood
the viscosity depends on the hematocrit. In a cylindrical blood vessel segment with a parabolic velocity profile and steady flow, equation 3.3 holds. However, at the entrance of a stenosis (Fig. 3.1,
site 2) the blood is accelerated fairly uniformly across the diameter,
implying high shear rates and high viscous losses close to the wall of
the stenosis. Within the stenosis the velocity profile gradually develops to a more parabolic velocity profile (Fig. 3.1, site 3) which is
associated with less viscous losses than at the entrance of the stenosis. Therefore, the pressure gradient is maximal at the entrance of
the stenosis and gradually decreases towards the end of it (see also
Fig. 2.7). Generally, a stenosis in a large artery is too short for
full development of a parabolic profile. So, equation 3.3 highly un-
derestimates the pressure drop Δp_v due to viscous friction, especially in a short stenosis.

The Bernoulli term is associated with acceleration and deceleration of blood at the sites of narrowing and widening of the stenosis,
respectively. Due to an increase of kinetic energy of the blood at
the site of narrowing, pressure inside the stenotic channel has
dropped. When decelerating flow at the site of widening is laminar,
this pressure drop is annihilated by a pressure increase, resulting
in a zero total pressure drop across the stenosis, due to the Bernoulli term. However, in a large artery the flow pattern distal to
a stenosis is generally highly turbulent, implying less pressure increase at the site of widening than pressure decrease at the site of
narrowing. At a high degree of turbulence the pressure drop across

a stenosis, associated with the Bernoulli effect, is described by

$$\Delta p_k \simeq \tfrac{1}{2}\rho \left(\frac{1}{A_v} - \frac{1}{A_s}\right)^2 q^2 \qquad 3.4$$

Thus, the total pressure drop Δp_s across a stenosis is the sum of the viscous term Δp_v and the Bernoulli term Δp_k (Clark, 1976).

$$\Delta p_s \simeq K_v \frac{128\,\pi\,\eta l_s}{A_s^2} + \tfrac{1}{2}\rho \left(\frac{1}{A_s} - \frac{1}{A_v}\right)^2 q^2 \qquad 3.5$$

where the constant $K_v > 1$, and K_v depends on the velocity profile within the stenosis.

A simplified method of calculating the pressure drop across a stenosis is provided in chapter 7.

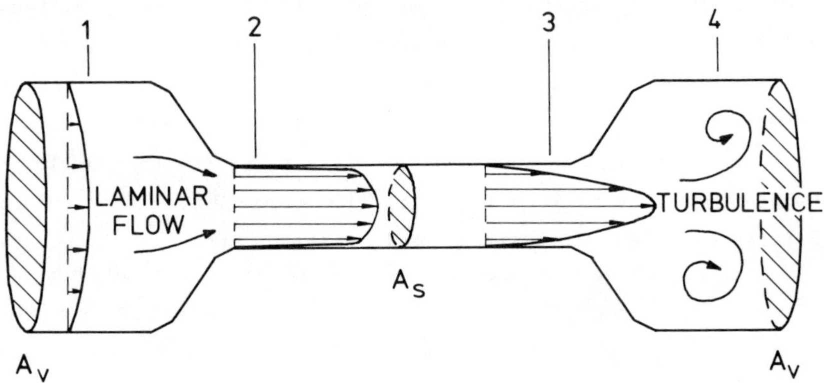

FIG. 3.1   Proximal to and within a stenosis flow is laminar. Distal to a stenosis turbulence often occurs, resulting in an extra pressure drop across the full length of the stenosis (between site 1 and 4). Within the stenosis an originally fairly flat velocity profile (site 2) develops gradually into a parabolic velocity profile (site 3).

3.2.3   Inertance (L) And Compliance (C)

Inertance and compliance, which together are called reactance, are two flow impedance properties which greatly affect the pulsatile flow waveforms in the arteries.

3.2.3.1 Inertance

Inertance is the most apparent impedance in normal large arteries where flow is pulsatile and exists primarily in the mass of the blood. It is analogous to electrical inductance symbolized as ⌇⌇⌇⌇⌇. Any change in electrical current through a coil of wire or an electrical inductance requires a peak of voltage difference in the direction of current flow to overcome the inductance of the coil. When the current is constant, voltage difference diminishes to a constant value. A more familiar example of the inertance property is that of pushing a person on an ice sled. One must push hard against the sled to accelerate it in order to get it going but after that, only enough push must be applied to overcome the small friction (resistance) of the runners. Similarly, any change of blood flow in the large blood vessels is attended with an acceleration transient in the pressure gradient to overcome the inertance of the blood mass.

The pressure drop due to an increase in kinetic energy in a vessel segment associated with inertance can be related to velocity of blood in the vessel segment as a non-stationary phenomenon in time. Then in a cylindrical vessel it holds:

$$\Delta p = \rho 1 \frac{dv}{dt} = \frac{\rho 1}{A_v} \frac{dq}{dt} = L \frac{dq}{dt} \qquad\qquad 3.6$$

The property of inertance in the aorta was first demonstrated by Spencer and Denison (1956) by direct measurement of the pressure difference along a 4 cm segment through which volumetric flow was simultaneously measured. Fig. 3.2 illustrates the experimental arrangement for measurement of two pressures simultaneously along the aorta. By subtracting the downstream pressure (p_2) from the upstream pressure (p_1), the total instantaneous pressure difference (Δp) across the segment was recorded (Spencer, 1960). The instantaneous blood flow was recorded with the square wave electromagnetic flowmeter applied between the pressure measurement points.

By inspection of Fig. 3.3, one can see that during the systolic upslope of the flow pulse, Δp reaches a maximum positive value and

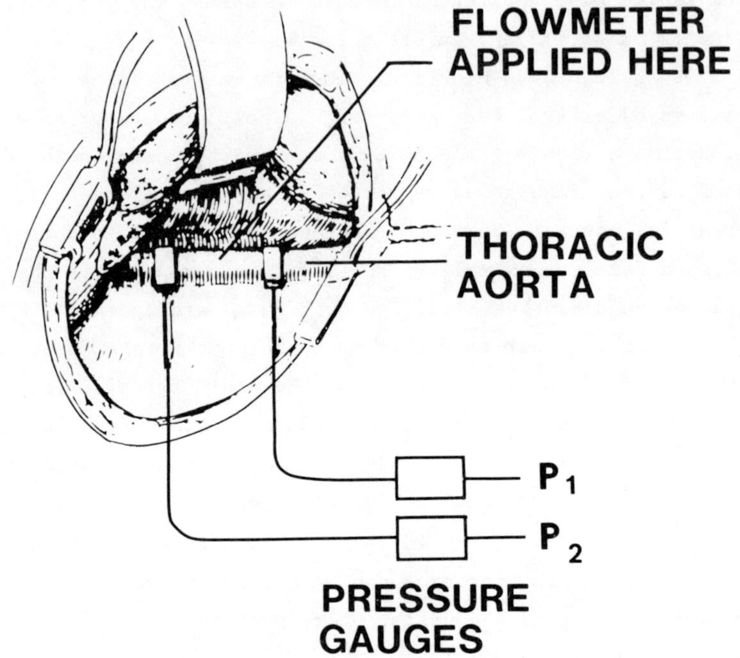

FIG. 3.2   How two pressure measurements are made
simultaneously along the aorta in order to record
Δp for Fig. 3.3.

during the following downstroke of flow, Δp reaches a maximum negative value. This phase lead of Δp compared to the flow pulse continues throughout subsequent oscillations in the heart cycle and is a
direct demonstration of the property of inertance in the large arteries. The flow pulse in Fig. 3.3 also demonstrates a marked backflow occurring at the end of systole. This oscillation is a prominent
feature of the arterial "resonant wave" which will be discussed
later.

Fig. 3.4 illustrates the property of inertance in the cardiac
ejection of the stroke volume into the ascending aorta (Spencer and
Greiss, 1962). The flow tracing represents the time course of normal
cardiac ejection as measured with the square-wave electromagnetic
flowmeter. Ejection begins with a positive Δp representing an

AORTIC FLOW AND DIFFERENTIAL PRESSURE

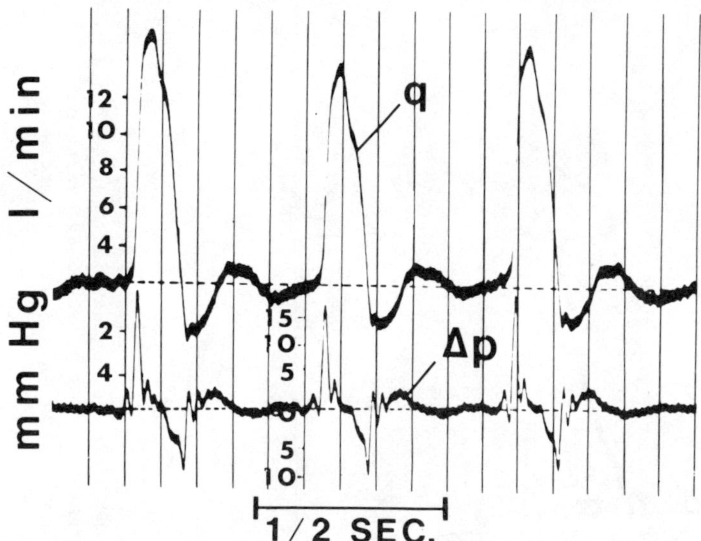

FIG. 3.3   Relationship between flow (q) through and
pressure difference (Δp) across a segment of the
descending thoracic aorta. A rise in flow is associated with a positive value of Δp due to acceleration of the blood in the vessel segment. Analo-
gously a decrease in flow is associated with a negative value of Δp. The two backflow phases,
visible in diastole, lag the oscillation in Δp by
approximately 90° as shown in Fig. 3.5 (1 mmHg =
0.133 kPa)

acceleration transient to open the valve and produce a steep systolic
upslope in the blood flow and velocity. When q reaches a peak at the
end of the Δp acceleration transient, Δp reverses direction. Through
the remainder of systole the blood is decelerated by a decrease of
ventricular pressure slightly less than aortic pressure until cardiac ejection is complete and ventricular contraction falls off. The
closure of the aortic valve is completed by a brief backflow transient whose reflection on the aortic pressure is called the "incisurd".

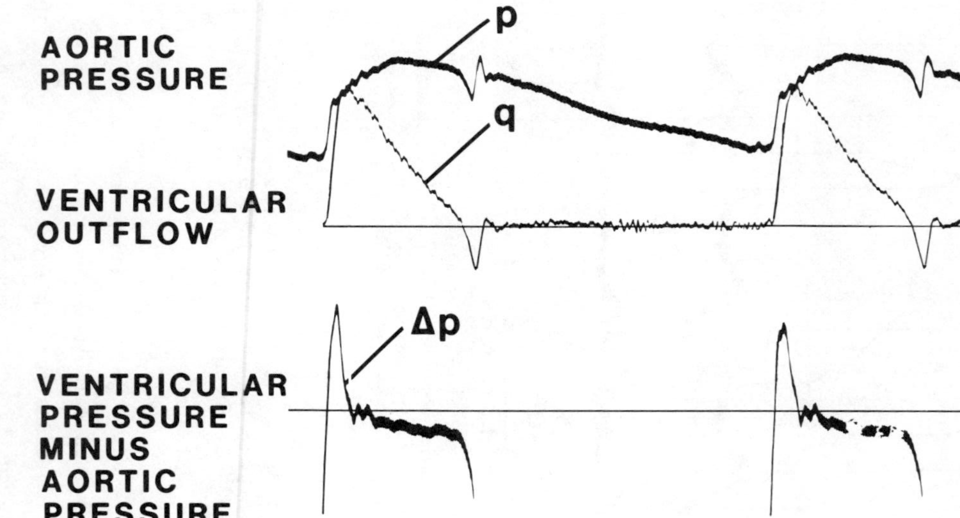

FIG. 3.4   The effect of inertia of the blood during cardiac ejection into the ascending aorta produces a large but brief positive acceleration pressure transient in early systole followed by a small but prolonged deceleration phase.

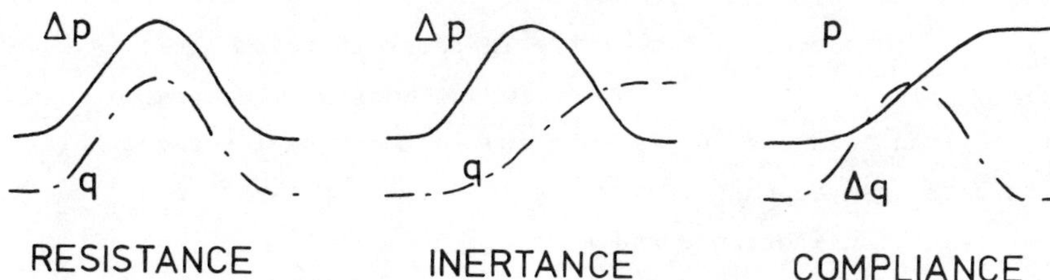

FIG. 3.5   Phase relation between pressure drop (Δp) and flow
(q) through a resistance and an inertance. In case of a
compliant channel the phase relation between pressure (p)
in the vessel and the difference between vessel inflow and
vessel outflow (Δq) is shown.

This ejection flow pulse is modified by the inertance and compliance
of the large arteries as it is transmitted along these arteries.

3.2.3.2   Compliance

Compliance is a property of the arterial wall arising from its distensibility which chiefly resides in the elastic fibers but also is
contributed to by smooth muscle and fibrous tissue. Vascular wall
compliance is analogous to electrical capacitance symbolized as ⊣⊢
Just as an electrical capacitance represents the ability to accumu-
late a charge of electricity and discharge it back into the same circuit, so compliance is the ability of the arteries to expand and
take up a volume of blood during systole, then discharge it along the
arteries during diastole.

   Compliance has been defined as

$$C = \frac{\Delta V}{\Delta p_c} \tag{3.7}$$

where ΔV is the change in volume of blood for any vessel segment and
Δp_c is the increase of the pressure inside the vessel segment.

   The property of compliance was expressed by the early "windkessel"
concept of the aorta which considered it to rapidly take up the
cardiac stroke volume during systole and slowly discharge it during

diastole through the peripheral resistance. Compliance of an artery varies directly with the 2nd power of the radius and directly with its length. In a vessel segment the increase in volume (ΔV) is equal to the difference (Δq) between inflow and outflow, integrated in time. Thus, Δq leads changes in pressure (p) in the vessel segment. Fig. 3.5 illustrates the phase relationship between pressure and flow for a resistance, an inertance and a compliance.

WINDKESSEL MODEL OF AORTA

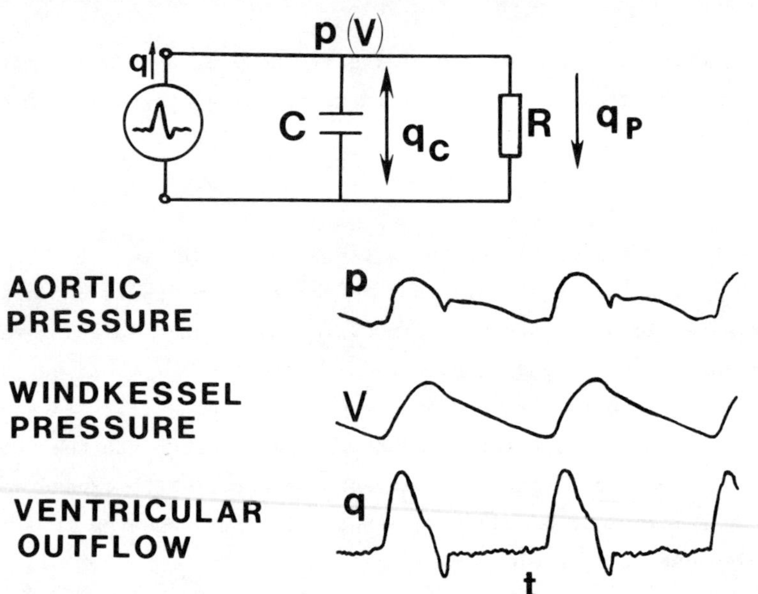

FIG. 3.6   The "windkessel" model of the arteries demonstrated as a simple R-C electrical circuit. It is not sufficient to explain the incisura and other details of the actually measured pressure seen in Fig. 3.4. q and p are measured, V is the voltage change computed in the analog model.

3.3   COMBINED ARTERIAL MODELS

Fig. 3.6 illustrates the windkessel concept as a simple electrical analogy. The heart is represented as a flow generator with an ejected flow pulse "q". The fraction of blood flowing in and out of the compliance (capacitator) C is designated q_C and that fraction flowing through the peripheral resistor R is designated q_R. The changes in the voltage (V) do not follow p closely because the inertial element (inductance) is not represented in the model.

This model of the arterial tree explains the gross rise and fall characteristic of the arterial pressure and shows how pressure falls as flow runs out through the peripheral resistance. This model is insufficient to explain other diagnostic characteristics of the peripheral arterial pressure and flow pulses. The windkessel model does not explain how there is a higher systolic pressure in the legs than in the arm. It does not explain backflow phases in the flow pulse and does not explain the dicrotic notch in the pressure pulse.

To explain these features a more complete and clinically useful model of the hemodynamics of the aorta and its major branches is illustrated in Fig. 3.7. Here the model of the arterial tree is expanded to two windkessels represented by two R-C circuits and connected by an inertial element, L. L represents the mass of blood in the aorta connecting the upper aorta and its branches with the abdominal aorta and its branches. The heart pumps directly into C_1, representing the compliance of the aortic arch and its immediate branches. This upper windkessel initially absorbs the left ventricular stroke ejection "bolus". In early systole, the inertance of blood in the descending aorta impedes flow into C_2, but after the acceleration transient passes, the flow pulse moves down the aorta to be absorbed by the C_2 windkessel. As the pressure in the lower windkessel then peaks out, the flow and pressure waves are reflected back up the aorta, producing a backflow phase throughout the descending aorta. According to this model, the arterial tree is a liquid filled elastic tube which is set into oscillation by each beat of the heart. The combination of the two capacitances with the inertance of blood mass lumped as a middle element produces resonant flow waves.

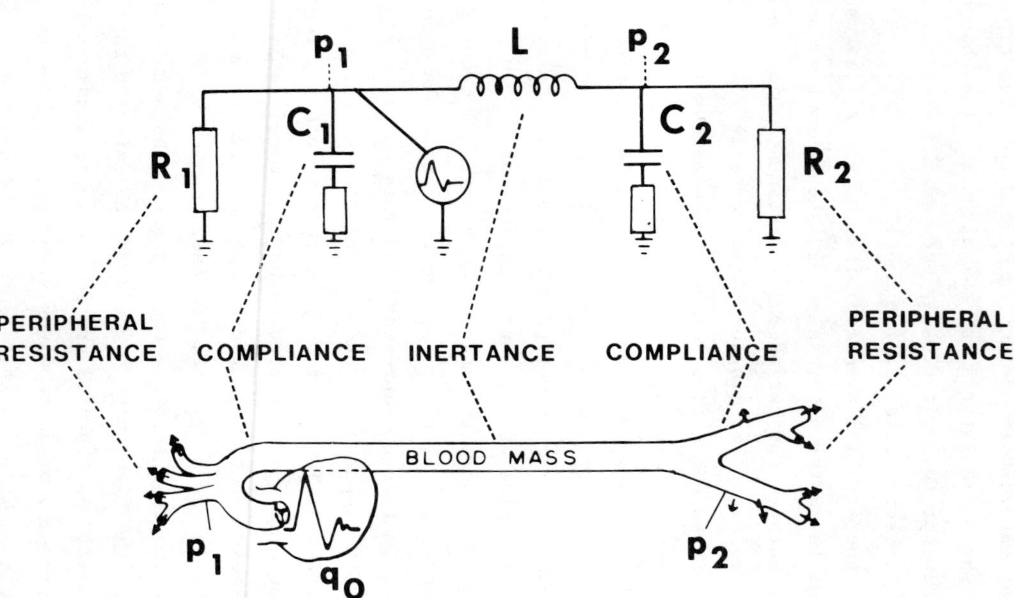

FIG. 3.7   A model of the hemodynamics of the aorta and its major branches which describes many clinically important features of flow and pressure pulses (Spencer and Denison, 1960).

AORTIC PRESSURE TRANSFORMATION

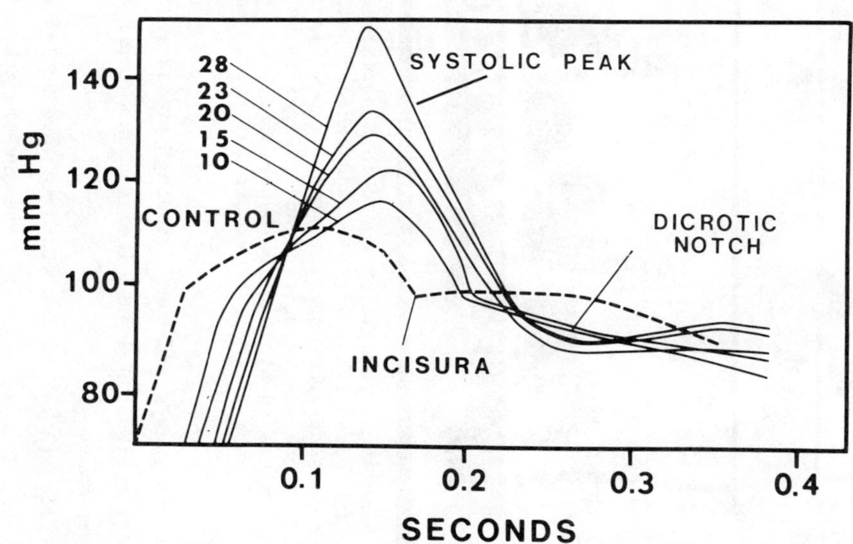

FIG. 3.8   Development of the standing wave in the pulse as it is transmitted to the lower aorta and femoral artery (according to Hamilton and Dow, 1939). 1 mmHg : 0.133 kPa.

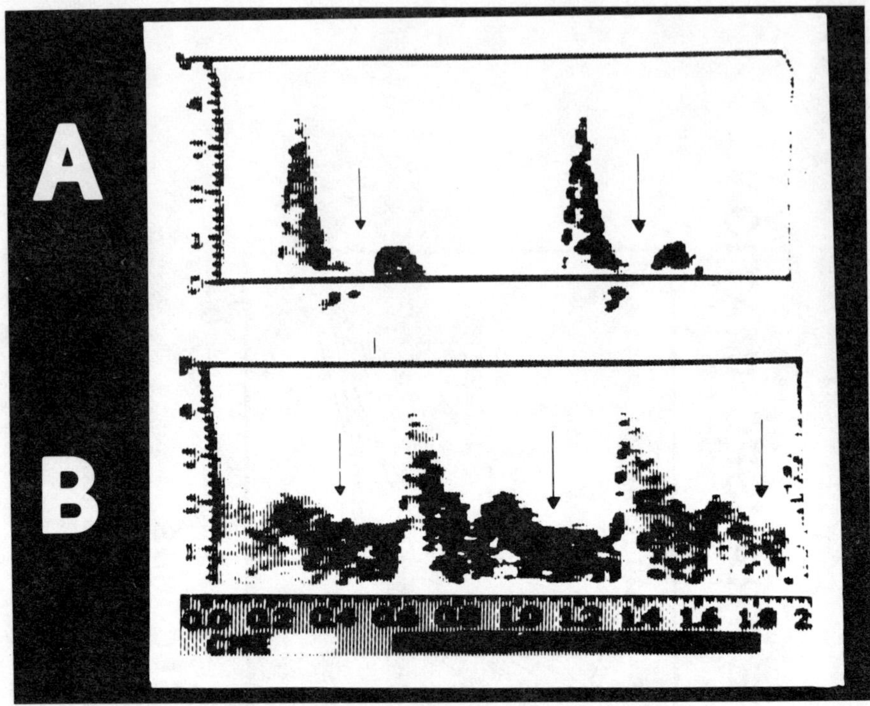

FIG. 3.9   Reactive hyperemia produced in a normal subject demonstrating increased diastolic flow and reduced pulsatility.
Panel A represents the normal resting velocities and panel B
represents increased velocities and changes in pulsatile characteristics caused by reactive hyperemia. These bi-directional
spectra were reproduced in black and white from color spectrum.

The resonant flow waves and the "standing" pressure waves of Hamilton
and Dow (1939) are part and parcel of the same interaction of inertance and compliance.

Fig. 3.8 illustrates how the standing wave builds up in the peripheral arteries. The peak systolic pressure in the femoral artery is
normally greater than that of the arms because C_2, of which the femoral artery is a part, is approximately 1/5 the size of C_1, of which
the brachial artery is a component. When the flow pulse enters the
smaller C_2 of the lower extremities, the pressure builds higher
during systole than it did when the flow pulse first entered C_1 of
the aortic arch and arms. Neglecting the influence of gravity the

pressures in the arms and legs, averaged over the heart cycle, are approximately the same because, as stated earlier, the large artery resistance produces only a very small mean pressure drop. The excess pressure in the femoral artery during systole is made up during diastole by the dicrotic notch pressure which is, at the time of its occurrence, lower than the upper aorta pressure.

3.4   PERIPHERAL FLOW PULSES

The resonant wave amplitude may be reduced by proximal or distal obstruction and by peripheral vasodilation. Vasodilation increases the flow and absorbs the resonant wave at the same time eliminating the backflow phase. These effects can be demonstrated non-invasively on a normal subject by listening to the Doppler sounds from the brachial artery while making a tight fist and releasing (Fig. 3.9). The resonant wave is apparent in the brachial artery of the relaxed arm. When the fist is clenched both forward and reverse flow are reduced because of reduced capacitance and increased resistance by muscular compression of the downstream arteries. Upon release of the clenched fist, reactive hyperemia in the distal bed produced increased flow and velocity in the brachial artery. Two features are particularly noticeable in the reactive hyperemia phase: (1) high diastolic flow with a decrease in the pulsatility and (2) elevation and diminution of the resonant wave. By observing the gradual return to the resting flow pattern, one can see on the spectral display and hear in the audio signals that diastolic backflow in the resting limb and the early diastolic dip in the vasodilation limb represent the same basic phenomenon, the resonant wave, superimposed on various levels of steady flow.

REFERENCES

Clark, C. (1976). The fluid mechanics of aortic stenosis. I. Theory and steady flow experiments. J. Biomechanics 9, 521-528.

Hamilton, W.F. and Dow, P. (1939). An experimental study of standing waves in the pulse propogated through the aorta. Am. J. Physiol. 125, 48-59.

Henquell, L., Lacelle, P.L., Honig, C.R. (1976). Capillary diameter
in rat heart in situ: relation to erythrocyte deformability, O_2
transport and O_2 gradients. Microvasc. Res. 12, 259-274.

Spencer, M.P. (1960). Differential pressure measurements: paired
transducer system. In Methods in medical research (Edited by Bruner)
pp. 340-346, Year Book publishers, Chicago.

Spencer, M.P. and Denison, A.B., Jr. (1956). Aortic flow pulse as
related to the differential pressure. Circ. Res. 4, 476-484.

Spencer, M.P. and Denison, A.B., Jr. (1960). An explanation of the
major features of the arterial pulse by means of a simple electronic analogy. Fed. Proc. 19, 87.

Spencer, M.P. and Denison, A.B., Jr. (1963). Pulsatile blood flow
in the vascular system. In Handbook of Physiology (2nd edition,
Edited by W.F. Hamilton), section 2, Circulation II. pp. 839-864,
American Physiological Society, Bethesda.

Spencer, M.P. and Greiss, F.C. (1962). Dynamics of ventricular
ejection. Circ. Res. 10, 274-279.

ADDITIONAL LITERATURE RELATED TO THE SUBJECT

Green, H.D. (1950). Circulatory system: physical principles. In
Medical physics (Edited by Glasser), volume 2. pp. 228-251,
Year Book publishers, Chicago.

McDonald, D.A. (1960). Blood flow in the arteries. William and
Wilkins, Baltimore.

Okino, H., Fujisaka, K., Sakaguchi, D., and Sasamoto, H. (1960).
Pulsatile blood flow in the arterial system. Respir. and Circ. 8,
49.

Peterson, L.H. (1954). The dynamics of pulsatile blood flow. Circ.
Res. 2, 127-139.

Spencer, M.P., Johnston, F.R., and Denison, A.B. (1958). Dynamics of
the normal aorta: "inertance" and "compliance" of the arterial
system which transforms the cardiac ejection pulse. Circ. Res. 6,
491-500.

Van der Tweel, L.H. (1957). Some physical aspects of blood pressure,
pulse wave and blood pressure measurements. Am. Heart J. 53, 4-17.

Warner, H.R. (1957). A study of the mechanism of pressure wave dis-
torsion by arterial walls using an electrical analogue. Circ. Res.
5, 79-84.

LIST OF SYMBOLS

Symbol	Unit	Description
A_s	m^2	cross-sectional area of stenosis
A_v	m^2	cross-sectional area of vessel
C	$m^3 Pa^{-1}$	compliance
d	m	vessel diameter
L	$Pa\ m^{-3} s^2$	inertance
l	m	length vessel segment
l_s	m	length stenotic channel
p	Pa	pressure
q	$m^3 s^{-1}$	volume flow
R	$Pa\ m^{-3} s$	flow resistance
t	s	time
v	$m\ s^{-1}$	velocity
Δp	Pa	pressure drop across vessel segment
Δp_k	Pa	Bernoulli term of pressure drop
Δp_s	Pa	pressure drop across stenosis
Δp_v	Pa	viscous term of pressure drop
Δq	$m^3 s^{-1}$	flow drop along vessel segment
Δv	$m\ s^{-1}$	velocity drop along vessel segment
η	$Pa\ s$	viscosity of blood
ρ	$kg\ m^{-3}$	density of blood

1 kPa = 7.5 mmHg; 1 mmHg = 133 Pa = 0.133 kPa.

CHAPTER 4
Doppler Ultrasound – Principle, Advantages and Limitations

R. S. Reneman *and* A. P. G. Hoeks

4.1   GENERAL INTRODUCTION

In this chapter basic aspects of Doppler ultrasound are described.
Attention is paid to such items as the Doppler principle, transmission of ultrasound – continuous wave versus pulsed Doppler
systems – signal processing, determination of blood flow velocity,
recording of velocity profiles and assessment of vessel diameter.
The advantages and limitations of Doppler devices are discussed.

Beside the determination of blood flow velocity and the assessment of vessel diameter, Doppler instruments are used to image the
circulation, allowing determination of the site of local blood flow
velocities. This technique is described as well.

The purpose of this chapter is to supply the user of this book
with some basic information to facilitate the reading of the following chapters.

4.2   PRINCIPLE

In Doppler flowmeters a beam of ultrasonic waves (at the MHz level)
is transmitted from a vibrating crystal diagonally through the vessel
wall into the blood stream. Some of the ultrasonic power is backscattered by the various structures in the body and received by
another or the same crystal. The crystals are mounted in a transducer (probe). Since ultrasound at the MHz level cannot be transmitted through air, acoustic gel is applied between the crystals and

78

the skin to improve acoustic coupling (Fig. 4.1).

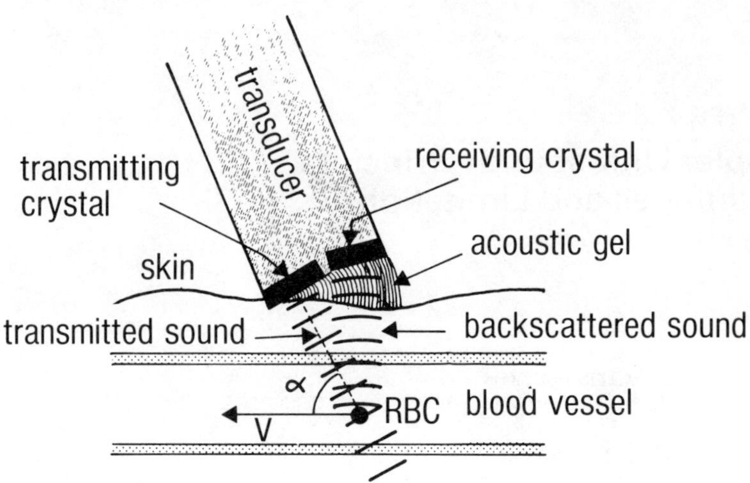

FIG. 4.1   The principle of continuous wave (CW) Doppler
flowmetry. The arrow indicates the flow velocity (v) direction. RB = red blood cell and α = the angle between
the transmitted sound beam and the direction of flow.
After Reneman et al, 1979.

Ultrasound backscattered from particles in the flowing blood,
mainly the red cells, is shifted in frequency by an amount proportional to the velocity of these particles. This frequency shift (the
Doppler shift = f_d), which is retrieved by mixing the transmitted
and received signals, is in the audio range and equals:

$$f_d = \frac{2f_t \, v \cos \alpha}{c} \qquad\qquad 4.1$$

in which f_t = transmitter frequency, v = velocity of the particles,
α = the angle between the transmitted sound beam and the direction of
the velocity of the particles (Fig. 4.1) and c = the velocity of
sound in the medium (1500 m/sec).

The Doppler signal does not contain one single frequency but a
spectrum of frequencies (the Doppler spectrum). The frequency distribution depends on such factors as unequal distribution of the red
blood cell velocity over the cross-sectional area of the vessel,
variations in the blood cell interspace and divergence and

non-uniformity of the sound beams, resulting in variations in α when the red blood cells are passing the ultrasonic beam (Peronneau et al, 1970).

The amount of power received at the crystals is determined by the amount of backscattering from the red blood cell plasma interface and the quantity of sound absorption by the tissues. Both the amount of backscattering and the quantity of absorption increase at higher emission frequencies (Wells, 1969). When the effects of reflection and absorption are combined, the backscattering from blood at a given distance from the transducer is strongest at a particular frequency (Reid and Baker, 1977). The higher the emission frequency the lower the penetration depth, but the better the ratio of power backscattered by particles in the blood and the power reflected by targets like vessel walls and tissue interphases. The higher the number of red blood cells moving in the ultrasonic beam, the higher the received Doppler power. The received signal contains power backscattered from the red blood cells as well as from the vessel wall. The signals induced by lateral wall motion are low in frequency, but high in amplitude. Their amplitude is approximately thirty times higher than that of the signals induced by moving red blood cells because the vessel wall-blood interface is a much better reflector than the red cell-blood interface, which causes mainly scattering.

4.3   SIGNAL PROCESSING

4.3.1 Introduction

In Doppler flowmeters the received signal is generally processed to an analog tracing, representing the average velocity as an instantaneous function of time (Fig. 4.2). In this technique valuable information, present in the Doppler signal, is ignored (Gosling, 1974).

An alternative way of analyzing received Doppler information is audio spectrum analysis. This processing technique, which yields detailed information about the Doppler signal, is more commonly used nowadays, especially because recently on-line audio spectrum analysis systems have become commercially available.

Some aspects of analog signal processing and audio spectrum

analysis, which might be of interest to clinicians, are discussed in
the present section.

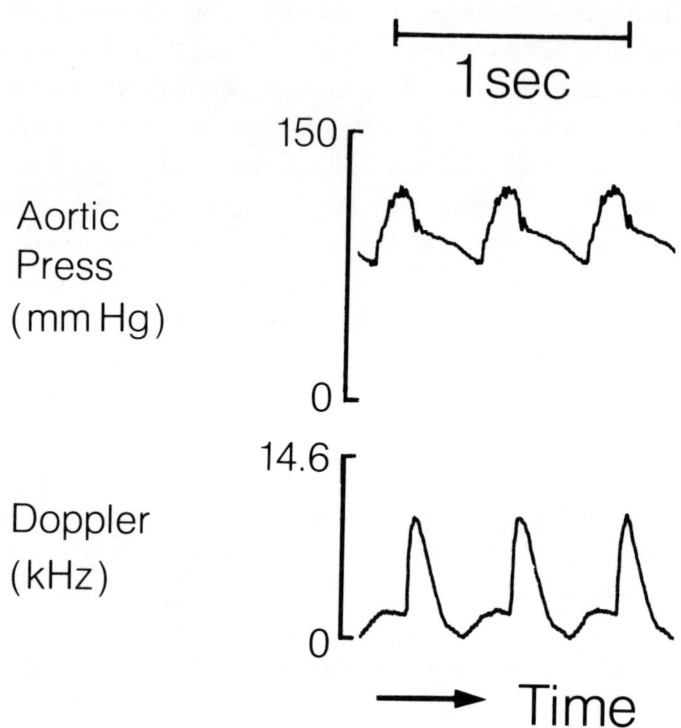

FIG. 4.2   The aortic pressure and instantaneous
femoral artery flow tracing as recorded with a CW
Doppler flowmeter, using a zero-crossing technique.
The analog Doppler tracing represents the average
velocity as an instantaneous function of time.

4.3.2 Analog Signal Processing

A simple method to retrieve the Doppler signal is mixing of the transmitted and received signals. This demodulation technique has the disadvantage that no information is obtained about the direction of the
blood flow (non-directional system). In directional systems a more
complicated demodulation technique is used (McLeod, 1967; Strandness
et al, 1969). In this technique the received signal is demodulated

with two signals at the transmitter frequency shifted 90° in phase with respect to each other, resulting in two Doppler signals which are 90° out of phase. The sign of the phase shift between both signals indicates the sign of the frequency shift, i.e. the direction of blood flow. This demodulation technique gives a bidirectional velocity output, but interference between forward and backward flow cannot be prevented. Separation of positive and negative Doppler shifts, and hence between forward and backward flow, can be achieved with single side-band demodulation. To diminish the influence of vessel wall motion signals and noise beyond the Doppler frequency band, the output of the demodulator is passed through a bandpass filter.

Determination of the mean frequency of the Doppler spectrum and conversion of the signal into an analog signal is usually performed with a zero-crossing meter. The output is an analog voltage proportional to the number of zero-crossings per unit of time. To diminish the counting of zero-crossings not related to blood flow velocity the comparator voltage of the Schmitt-trigger is set at such a level that the zero-crossing meter will not be activated by low level signals (e.g. noise). The analog tracing obtained with this technique theoretically gives the average velocity as an instantaneous function of time.

Difficulties encountered in this signal processing technique were examined (Reneman et al, 1973; Reneman and Spencer, 1974) and discussed (Reneman and Hoeks, 1977) previously. The zero-crossing meter does not exactly measure the mean frequency of the Doppler spectrum, corresponding to the average velocity, but a value higher than the mean Doppler frequency (Rice, 1944; Peronneau et al, 1970). This systematic error depends on the shape and the width of the Doppler spectrum; the broader the spectrum with respect to the average frequency the larger the error. The zero-crossing meter appears to be accurate only for a single frequency or a relatively narrow frequency spectrum.

An additional disadvantage of zero-crossing meters is that the output is only independant of the audio signal level if this level exceeds the noise level by at least 10 times (Reneman and Spencer, 1974), provided that the threshold of the zero-crossing detector is

adjusted, so that random fluctuations due to noise are just eliminated. The latter is required to prevent shifting of the instantaneous velocity tracing from the zero-line because, in cerebral vascular disease valuable information can be derived from the systolic
and diastolic amplitudes of the analog velocity tracings of the
common carotid artery (see Chapter 6). Another disadvantage is that
the output of the zero-crossing meter becomes unreliable if high
amplitude low frequency components, induced by vessel wall motion,
are present in the Doppler signal (Reneman and Spencer, 1974). To
reduce the noise contribution and vessel wall motion artefacts, a
bandpass filter is used which limits the frequency range of the audio
signal albeit at the cost of low and high velocity information. The
upper frequency limitation can be a problem, especially in stenotic
regions.

Recently an alternative method has been described to determine the
mean velocity from the Doppler spectrum (Arts and Roevros, 1972; Reid
et al, 1974; Angelsen, 1975). This method looks promising because it
properly takes into account the spectral distribution of frequencies.

In spite of this possible improvement in the processing of analog
tracings, the use of these tracings has limitations. Although shape
analysis and the systolic and diastolic amplitudes of the analog
velocity tracings do give important information about the site and
severity of arterial stenosis (see for review, Reneman et al, 1979),
valuable information, present in the Doppler signal, is eliminated
in this technique.

4.3.3 <u>Audio Spectrum Analysis</u>

Detailed information can be obtained from the Doppler signal when
audio spectrum analysis is used (Baskett et al, 1977; Blackshear et
al, 1979). In audio spectrum analysis sonograms are produced in which
frequencies are given as an instantaneous function of time and the
intensity of the pattern represents the amplitude of the frequencies,
indicating the number of red blood cells moving at a given velocity
(Fig. 4.3). Sonograms yield information about the maximum blood flow
velocity and the velocity distribution of the red blood cells, giving

insight in the flow pattern (Reneman and Spencer, 1979). These are
important variables in evaluating the peripheral arterial circulation.
The maximum velocity can, for example, be used to diagnose lesions in
arteries (Gosling and King, 1974) or to estimate the degree of arterial stenosis (Spencer and Reid, 1979).

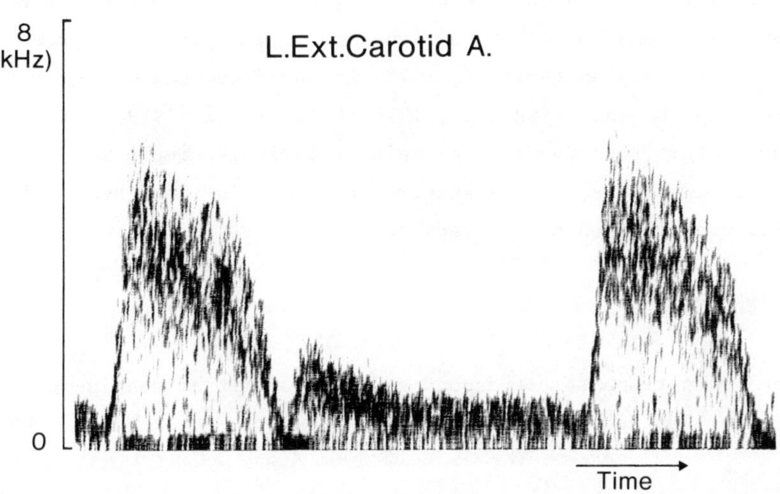

FIG. 4.3   Sonogram as recorded in the left external carotid
artery halfway the bifurcation and the mandible with a CW
Doppler device in a volunteer without symptoms of cardiovascular disease. The outline of the sonogram is regular.
Frequencies tend to concentrate near the maximum frequency, suggesting that in normal arteries red blood cells
tend to travel near the maximum velocity, which is representative of plug flow.   After Reneman et al, 1979.

In laminar flow, the outline of the sonogram, defined as the line
following the maximum frequencies during the cardiac cycle, is regular (Fig. 4.3). In turbulent flow the spectrum broadens and the outline of the sonogram becomes irregular due to the random changes in
flow velocities occurring at any time during the cardiac cycle (Figs.
4.10 and 4.11).

Although audio spectrum analysis yields valuable information, this
processing technique is not commonly used in clinical practice. Mainly because, until recently, audio spectrum analysis could only be

performed off-line, which is a time-consuming procedure and unsuitable for routine clinical applications. Besides, the mean frequency of the Doppler spectrum is difficult to derive in off-line analysis techniques. At present, however, several on-line systems are commercially available. Beside the on-line presentation of sonograms, these systems offer the possibility of determining both the maximum and the mean velocity as an instantaneous function of time. To obtain the maximum velocity as a continuous function of time either frequency analysis (Skidmore and Follett, 1978) or dedicated analog processing equipment can be used (Angelsen, 1976; Hatle et al, 1979). The latter approach offers the advantage of being relatively simple but the detailed information about the spectral distribution, and hence the velocity distribution of the red blood cells, is lost.

4.4   EMISSION OF ULTRASOUND

In Doppler instruments ultrasound can be transmitted continuously (CW Doppler) or intermittently (pulsed Doppler).

4.4.1 CW Doppler

In CW Doppler instruments the ultrasonic beam is usually transmitted from one crystal and the backscattered ultrasound received by another one (Fig. 4.1). These systems are easy to build and to operate, but vessel wall motion artefacts (see section 4.2) are often difficult to eliminate, at least without simultaneously eliminating the signals from blood flowing at low velocities. The high amplitude low frequency signals due to lateral wall motion influence the shape of the frequency spectrum and, therefore, the accuracy of the mean velocity determination with the zero-crossing meter. Vessel wall motion signals can even mask the presence of high velocity information, necessitating additional attenuation at the lower end of the bandpass filter. An increase in roll-off rather than increasing the cut-off frequency is usually sufficient and preferable. This phenomenon is seen in analog signal processing with zero-crossing meters (Reneman and Spencer, 1974) as well as in audio spectrum analysis (Fig. 4.4).

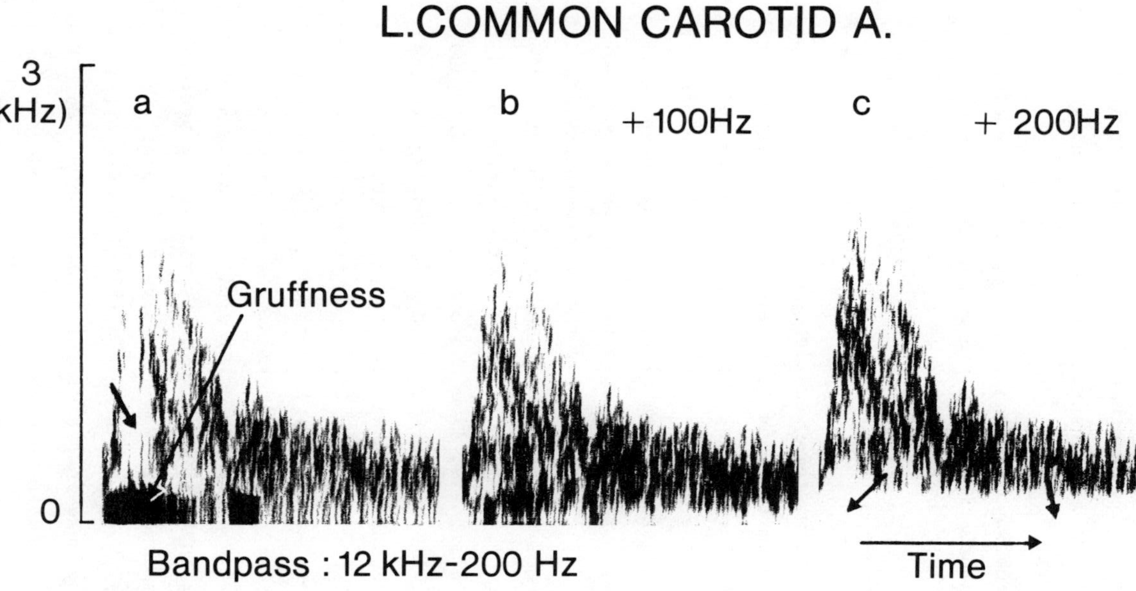

FIG. 4.4   Panel a: loss of high frequency information (see arrow) due to high amplitude low frequencies. The sound produced by the latter frequencies is gruff. Panel b: additional filtering at the lower end of the bandpass attenuates the high amplitude low frequencies and reveals the presence of high frequencies. Panel c: too much filtering leads to loss of flow velocity induced frequencies (see arrow). Sonograms recorded with CW Doppler instrument. After Reneman and Spencer, 1979.

86

As shown in section 3.2, the zero-crossing meter is only accurate
for a single frequency or a narrow frequency spectrum. This limits
the applicability of zero-crossing meters in combination with CW
Doppler flowmeters, where a wide spectrum is fed into the meter. This
wide spectrum results among other things from the variations in velocity of the red blood cells over the cross-sectional area of the
blood vessel.

In transcutaneous applications, CW Doppler instruments do not
yield accurate information about the vessel diameter.

4.4.2 Pulsed Doppler

In pulsed Doppler flowmeters usually one single crystal, operating
alternately as transmitter and receiver, is used. The crystal receives the backscattered signals from the red blood cells and the
vessel wall during the interval between pulses. An electronic gate
allows selection of scatterings either from the vessel wall or the
red blood cells at a given distance from the transducer (Fig. 4.5).
This makes it possible to determine the mean velocity in a small
sample volume as an instantaneous function of time at various sites
in an artery (Fig. 4.6), thus avoiding contamination of the desired
signal by unwanted signals, like those from vessel wall and veins.

In principle, with single-channel pulsed Doppler systems, the
velocity profile – that is the velocity distribution over the cross-
sectional area of the vessel – can be determined. In these systems,
however, during one cardiac cycle the velocity as an instantaneous
function of time can only be determined at one site in the vessel.
Therefore, synthesis of the velocity profiles during a cardiac cycle
requires that the instantaneous velocity signals at various sites in
an artery are assessed during consecutive heart beats. This limits
the applicability of these systems in the diagnosis of peripheral
artery diseases because in the vicinity of stenotic lesions the velocity profile was found to change locally during one cardiac cycle
(Wille, 1979). Besides, the positioning and maintenance of the sample
volume at the site of interest requires some skill. The simultaneous
use of pulsed echo (e.g. B-mode imaging of the vessel wall or two-

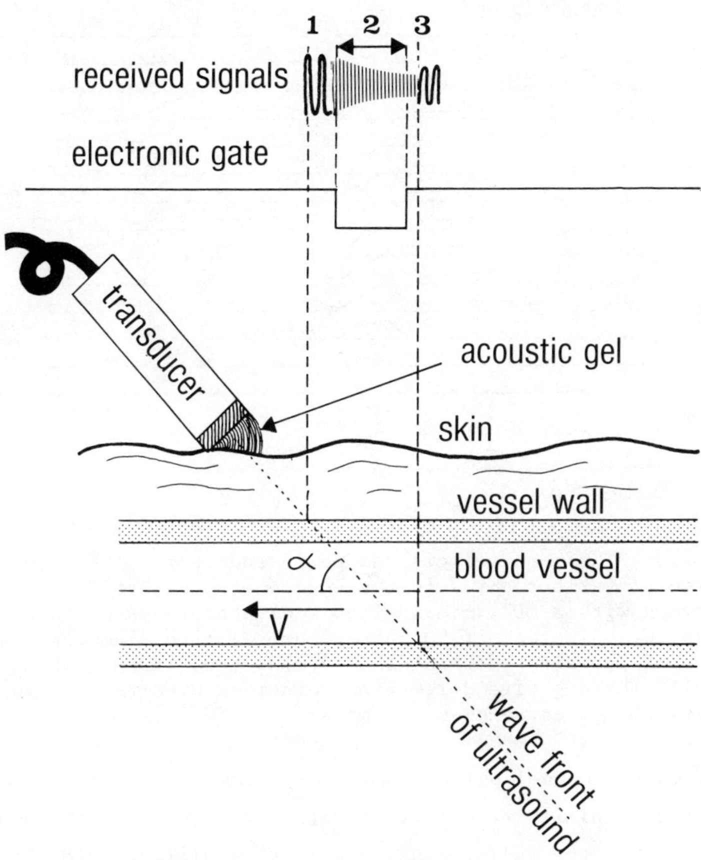

FIG. 4.5   The principle of single channel pulsed
Doppler flowmetry. The arrow indicates the flow
direction. 1 and 3 = ultrasonic power reflected
from the anterior and posterior wall of the
vessel, respectively. 2 = ultrasonic power backscattered from the red blood cells. The electronic gate is so adjusted that mainly the
signal which contains the flow velocity
information is processed. After Reneman and
Spencer, 1979.

dimensional imaging) and pulsed Doppler systems (Barber et al, 1974;
Pourcelot, 1977) will facilitate this task, but reduces the maximum
detectable frequency which is limited in pulsed Doppler devices
anyhow (see below).

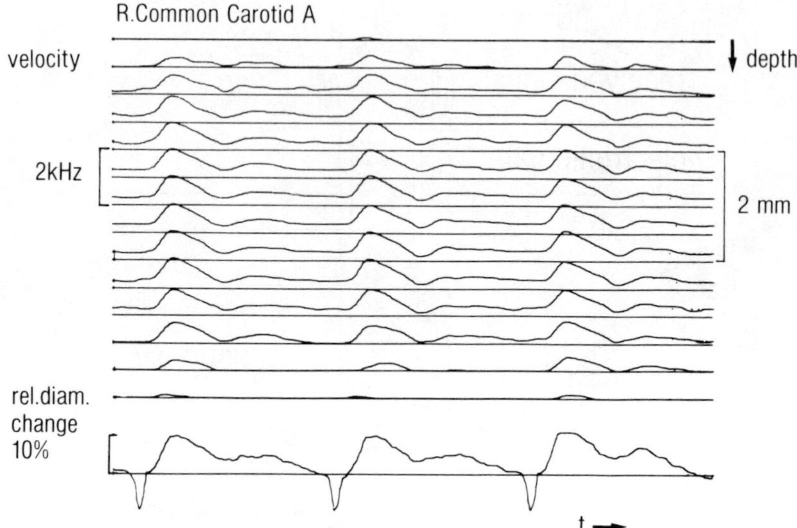

FIG. 4.6   The mean velocity as an instantaneous function of
    time at various sites in the common carotid artery as re-
    corded with a multi-channel pulsed Doppler system in a
    healthy volunteer of 20 years. The relative diameter chan-
    ges of the artery during the cardiac cycle are shown as
    well. The negative deflection coincides with the R-wave of
    the ECG. After Reneman, in press.

Recently multi-channel pulsed Doppler systems have been developed
that have the ability to detect simultaneously and instantaneously
velocities over the full range of interest (Anliker, 1978; Brandestini, 1978; Hoeks et al, 1981). With these systems the velocity
profile can be recorded on-line at discrete time intervals during
one cardiac cycle (Fig. 4.7). To obtain reliable velocity profiles,
the sample resolution has to be high and the sample distance along
the ultrasonic beam must be small. A limited number of independent
sample points along the cross-section of the vessel provides more
parabolic velocity profiles and significantly overestimates vessel
diameter (Anliker, 1978). Small sample volumes can only be obtained
if the effective duration of the measurement is small and the beamwidth is narrow. The effective duration is set by the duration of
emission combined with the bandwidth of the receiver section and the

gate-width (Peronneau et al, 1974). Increasing the bandwidth and shortening the duration of emission (high emission frequency) and the gate-width will reduce the sample volume, but will decrease the signal-to-noise ratio. Besides, the relative changes in artery diameter during the cardiac cycle can be obtained on-line (Fig. 4.7) by means of these systems (Hoeks et al, 1980; Reneman, in press).

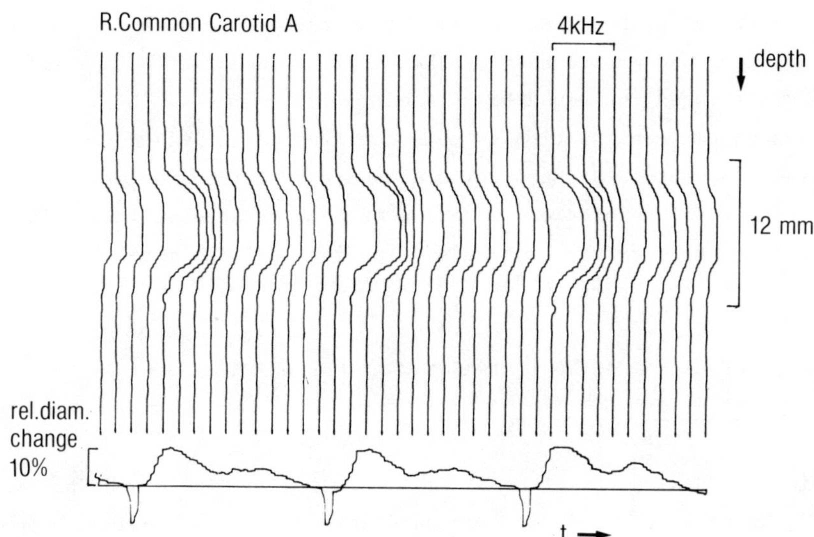

FIG. 4.7   The velocity profile in the common carotid artery
    at discrete time intervals during the cardiac cycle as syn-
    thetized from the analog velocity tracings shown in Fig.
    4.6. The relative diameter changes of the artery during the
    cardiac cycle are shown as well. The negative deflection co-
    incides with the R-wave of the ECG. After Reneman, in press.

An additional advantage of pulsed devices is that small volume samples are taken along the vessel diameter, so that a narrow frequency spectrum is fed into the zero-crossing meter. Hence the error made in determining the mean frequency of the Doppler spectrum with this meter is small in pulsed systems.

Although the advantages of pulsed Doppler systems are obvious, it should be noticed that in these devices problems are encountered which are not met in CW systems. The circuitry of pulsed devices is more complex and the Doppler bandwidth of the system is limited. The maximum Doppler velocity that can be detected unambiguously depends on the distance between transducer and vessel and the transmitter frequency (Reneman and Hoeks, 1977; Hoeks et al, 1979). The distance sets an upper bound to the pulse repetition frequency, which should exceed the maximum Doppler frequency at least twice. The limited bandwidth of pulsed Doppler devices is especially a problem when maximum velocities have to be recorded within tight arterial stenoses. An additional problem encountered under these circumstances is the positioning and maintenance of the sample volume within the stenosis, especially when the sound beam is narrow.

Transducer construction is more complicated in pulsed than in CW systems. To obtain a sharp pulse of short duration, necessary for adequate axial resolution, matching of the characteristic impedance of the crystals to the backing and loading media is more critical than in CW flowmeters.

4.5 VELOCITY IMAGING (ULTRASONIC ARTERIOGRAPHY)
4.5.1 <u>Principle</u>

Beside the assessment of blood flow velocity, CW (Reid and Spencer, 1972; Spencer et al, 1974; Curry and White, 1978) and pulsed Doppler (Mozersky et al, 1972; Fish, 1975) systems are used to image the circulation. In this technique a focussed transducer of a directional Doppler flow system is connected to a mechanical scanning arm. The position of the transducer and beam is electronically sensed by position sensing circuitry which causes the beam of an image storage oscilloscope to move in correspondence with the position of the transducer. The output of the directional Doppler flow system intensifies the Z-axis of the image storage oscilloscope only for a given direction of flow, eliminating venous flow signals superimposed on the arterial ones (especially important for CW devices). To improve the virtual resolution the threshold of the Z-axis beam control circuit is set so that only signals above, for instance, 10 per cent of peak flow will be imaged. By passing repeatedly the transducer over the artery and following the artery along its course a two-

dimensional picture of arteries can be made. The image formed is
similar to the anatomical display of X-ray arteriography, but represents a functional projection of local blood flow velocities. The
sound beam of the transducer generally makes an angle of approximately 60° with the longitudinal axis of the body to obtain an adequate
Doppler signal from the flowing blood. This angle should be maintained during the scanning procedure.

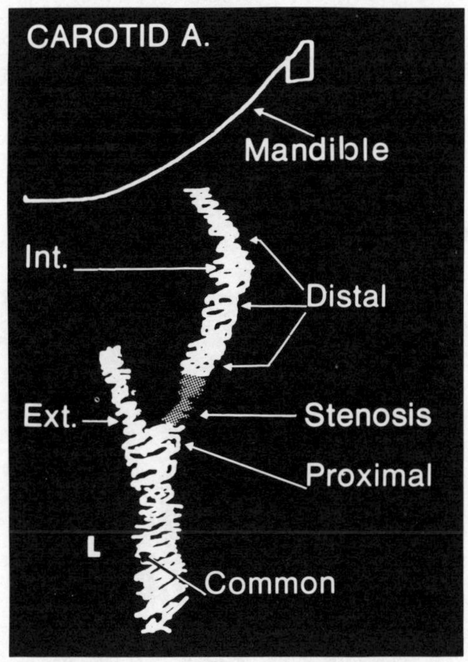

FIG. 4.8   Cervical carotid image
with sites of recording of the
audio spectrum in a patient with
an internal carotid artery stenosis, using a CW Doppler instrument. After Reneman and Spencer,
1979.

In Fig. 4.8 a cervical carotid image of a patient with stenosis
of the internal carotid artery is shown. The site of stenosis is
localized by the local shift of the sound to a higher frequency
domain. Additional information is obtained from the local narrowing

of the carotid image, usually seen at the site of stenosis. The image shown in Fig. 4.8 is recorded with a 5 MHz CW Doppler flow system, using a transducer which focusses at 2.5 cm to a minimum beam diameter of 2 mm.

Imaging with CW Doppler systems requires a narrow beamwidth to enhance spatial resolution. Unlike CW Doppler devices, pulsed Doppler systems can image the cross-section of arteries (Fish, 1975).

Although attempts have been made to estimate the degree of carotid artery narrowing from ultrasonic velocity images (Barnes et al, 1976), this imaging technique is generally used to localize properly the site of velocity recording in relation to the position of the steno-sis.

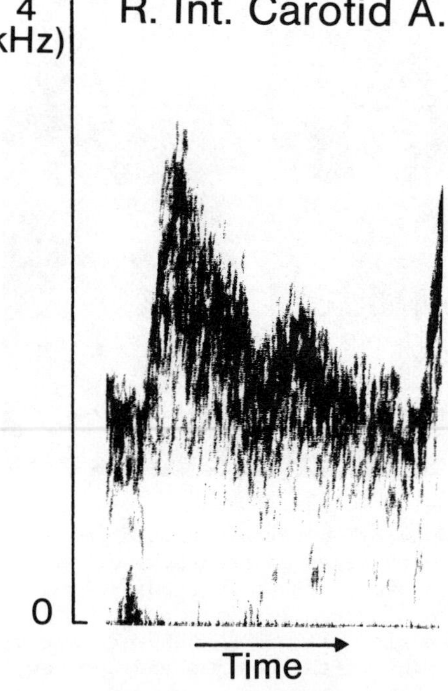

FIG. 4.9   Sonogram as recorded in the internal carotid artery with a CW Doppler instrument in a volunteer without symptoms of cardiovascular disease.

4.5.2 <u>The Combination Of Audio Spectrum Analysis And Velocity Imaging</u>

By combining a CW Doppler flowmeter with imaging system and audio spectrum analysis rather detailed information can be obtained about the flow disturbances along stenosed arteries (Reneman and Spencer, 1979).

In the internal carotid artery, for instance, the normally present plug flow (Fig. 4.9), giving rise to a smooth sound, changes to more parabolic flow within a stenosis (65-74%) and to turbulent flow distal to a stenosis (Fig. 4.10).

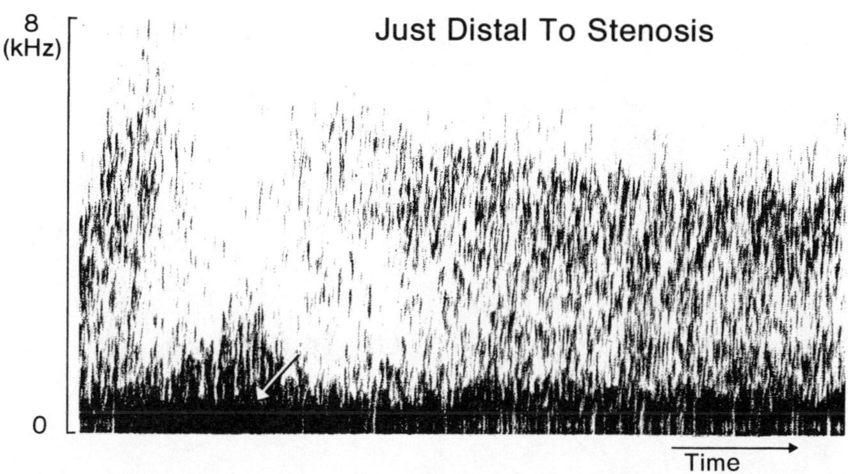

FIG. 4.10   Sonogram as recorded in the internal carotid artery just distal to a stenosis with a CW Doppler instrument. Note gruffness (see arrow). After Reneman and Spencer, 1979.

The maximum velocity increases significantly at these sites. In plug flow (= flat flow profile), the red blood cells tend to travel near the maximum velocity, giving rise to a concentration of frequencies near the maximum frequency. In parabolic flow, the red blood cells are travelling at various velocities, resulting in widening of the spectrum and a more or less even distribution of frequencies over the spectrum. Plug flow and parabolic flow are laminar flow patterns.

94

The presence of turbulence distal to the stenosis is indicated by
the widening of the spectrum, the loss of concentration of the frequencies near the maximum and an irregular outline of the sonogram.

Just distal to a stenosis the high velocity turbulence induces
lateral wall vibration, producing a gruff quality sound which simulates the bruit heard with the stethoscope. In the sonogram lateral
wall vibration appears as high amplitude low frequencies, less than
675 Hz for a 5 MHz Doppler instrument, usually throughout systole
(Fig. 4.10). High amplitude low frequency components are occasionally seen in sonograms of normal carotid arteries. In normal arteries,
however, these low frequencies are present in systole only during
acceleration and deceleration of blood flow.

The gruff quality sound and the high amplitude low frequencies
disappear approximately 2 cm downstream to the stenosis. At this
site the maximum frequencies of the sonogram decrease significantly
as compared with just distal to the stenosis, but all signs of turbulence persist (Fig. 4.11). The low velocity turbulence produces
at this site a boiling or fluttering sound.

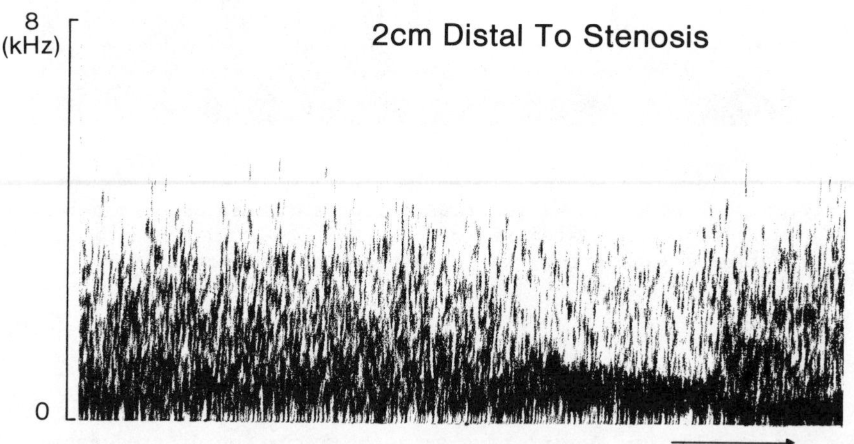

FIG. 4.11   Sonogram as recorded in the internal carotid artery
approximately two cm distal to a stenosis with a CW Doppler
instrument. After Reneman and Spencer, 1979.

Three to 4 cm downstream to a stenosis flow has become laminar
again as indicated by the generally regular outline of the sonogram
and the tendency of the frequencies to concentrate near the maximum
frequency (Reneman and Spencer, 1979).

Since disturbances in the flow pattern along narrowed arteries do
occur at relatively slight degrees of stenosis (Barnes et al, 1976;
Giddens et al, 1976; Sandmann et al, 1978) the detection of these
disturbances contributes to the early diagnosis of carotid artery
disease (see Chapter 9).

4.6   ADVANTAGES AND LIMITATIONS

The major advantage of the Doppler technique is that it is non-invasive. In humans mean and maximum blood flow velocity can be recorded
transcutaneously as an instantaneous function of time in arteries
and veins. When using audio spectrum analysis CW Doppler instruments
do give insight in the flow pattern in arteries. More detailed information about flow pattern can be obtained with multi-channel pulsed
Doppler systems, which allow the on-line recording of velocity profiles at discrete time intervals during one cardiac cycle. Moreover,
Doppler flowmeters are used to image the circulation. Imaging of the
circulation allows the determination of the site of local blood flow
velocities, while the site of artery stenosis can be determined with
this technique. Additional advantages are that with directional
Doppler systems, using separated audio channels for positive and
negative frequencies, forward and backward flow can be recorded inde-
pendently so that the existence of backward flow can be assessed when
the net flow direction is antegrade and contaminating venous flow
signals can be recognized when measuring arterial blood flow velocity.
A zero flow reference can easily be obtained by disconnecting the
input to the signal processing system. Doppler systems can easily
be used in radio telemetry.

In Doppler flowmeters quantitative flow velocity information is
still difficult to obtain, mainly because the angle α (Eq. 4.1 and
Fig. 4.1) cannot be determined accurately and precise assessment of
the average frequency of the Doppler spectrum is difficult to achieve.

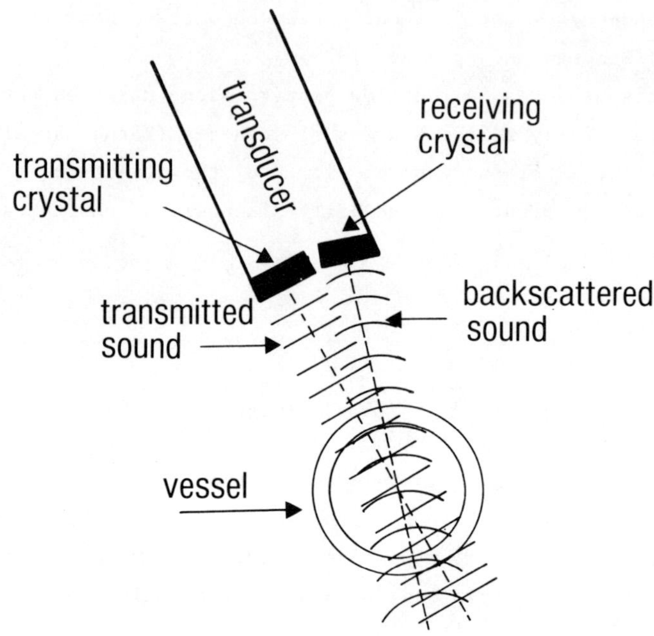

FIG. 4.12   Schematical representation of one of
the possible errors in Doppler flowmetry. The
whole cross-sectional area of the artery is
not bathed in the ultrasonic beam so that flow
velocity is averaged along the diameter rather
than over the cross-sectional area of the
vessel. This leads to overestimation of the
mean flow velocity over the cross-sectional
area (see text). After Reneman et al, 1979.

Hence determination of the average velocity over the cross-sectional
area of the blood vessel is subject to errors.

In Doppler devices often the whole cross-section of the vessel is
not bathed in the ultrasonic beam so that in CW systems flow velocity is averaged along the diameter rather than over the cross-sectional area of the vessel (Fig. 4.12). Averaging along the diameter
of the femoral artery, for instance, can easily overestimate the
mean velocity over the cross-sectional area by some 33 per cent
(Gessner, 1969).

The accuracy of transcutaneous vessel diameter determination with pulsed Doppler devices is limited and depends on the emission duration and the width of the gate. The accuracy with which the site of the vessel wall can be determined is usually limited to one mm, which means a maximum error of 25 per cent in determining the diameter of a vessel of 8 mm. In pulsed echo devices, in which the sound beam is approximately perpendicular to the axis of the artery, the diameter of the vessel can be determined more accurately.

At present reliable transcutaneous volume flow measurements in small arteries cannot be made with Doppler flowmeters, because of the inaccuracy in determining the average velocity over the cross-sectional area of the blood vessel and vessel diameter. Changes in the diameter of arteries during the cardiac cycle can be assessed more accurately. Even with CW Doppler devices arterial wall displacements of 25 μm can be distinguished under certain conditions (Olsen, 1977).

Because of their non-invasiveness Doppler techniques are especially suitable for studies requiring repeated measurements, for example, studies to evaluate the results of surgical or non-surgical treatment, or epidemiological studies.

Although quantitative flow velocity information is still difficult to obtain and volume flow cannot be determined accurately, Doppler instruments have been shown to be an asset in the diagnosis of carotid artery disease. The usefulness of the techniques developed to evaluate the functional state of the cerebral circulation will be discussed in the following chapters.

REFERENCES

Angelsen, B.A.J. (1975). Transcutaneous measurement of aortic blood velocity by ultrasound; a theoretical and experimental approach. PhD-thesis no. 75-78W, N.T.H., Trondheim, Norway.

Angelsen, B.A.J. (1976). Analog estimation of the maximum frequency of Doppler spectra in ultrasonic blood velocity measurements. Report 76-21-W. Div. of Eng. Cybernetics, N.T.H., Trondheim, Norway.

Anliker, M. (1978). Diagnostic analysis of arterial flow pulses in man, In Cardiovascular system dynamics (Edited by J. Baan, A. Noordergraaf and J. Raines). pp. 113-123, MIT Press, Cambridge

Arts, M.G.J. and Roevros, J.M.J.G. (1972). On the instantaneous measurement of blood-flow by ultrasonic means. Med. Biol. Engng. 10,
23-34.

Barber, F.E., Baker, D.W., Strandness, D.E., and Mahler, G.D. (1974).
Duplex scanner II. Ultrason. Symp. Proc. IEEE, Cat. = 74 CHO 8961
Transactions on Sonics and Ultrasonics.

Barnes, R.W., Bone, G.E., Reinertson, J., Slaymaker, E.E., Hokanson,
D.E., and Strandness, D.E. (1976). Non-invasive ultrasonic carotid
angiography: prospective validation by contrast arteriography.
Surgery 80, 328-335.

Baskett, J.J., Beasley, M.G., Murphy, G.J., Hyams, D.E., and
Gosling, R.G. (1977). Screening for carotid junction disease by
spectral analysis of Doppler signals. Cardiovasc. Res. 11,
147-155.

Blackshear, W.M., Philips, D.J., Thiele, B.L., Hirsch, J.H., Chikos,
P.M., Marinelli, M.R., Ward, K.J., and Strandness, D.E. (1979).
Detection of carotid occlusive disease by ultrasonic imaging and
pulsed Doppler spectrum analysis. Surgery 86, 698-706.

Brandestini, M. (1978). Topoflow - A digital full range Doppler
velocity meter. IEEE Transactions on Sonics and Ultrasonics SU-25,
287-293.

Curry, G.R. and White, D.N. (1978). Color coded ultrasonic differential velocity arterial scanner (Echoflow). Ultrasound Med. Biol. 4,
27-35.

Fish, P.J. (1975). Multichannel, direction-resolving Doppler angiography. Excerpta Medica International Congress Series, 363, pp. 153-
159. Excerpta Medica, Amsterdam.

Gessner, U. (1969). The performance of the ultrasonic flowmeter in
complex velocity profiles. IEEE Transactions Biomedical Engineering BME-16, 139-142.

Giddens, D.P., Mabon, R.F., and Cassanova, R.A. (1976). Measurements
of disordered flows distal to subtotal vascular stenoses in the
thoracic aortas of dogs. Circ. Res. 39, 112-119.

Gosling, R.G. (1974). General discussion: the usefullness of zero-
crossing meters. In Cardiovascular applications of ultrasound
(Edited by R.S. Reneman). pp. 455-456, North-Holland/American
Elsevier, Amsterdam-London-New York.

Gosling, R.G. and King, D.H. (1974). Continuous wave ultrasound as
an alternative and complement to X-rays in vascular examinations.
In Cardiovascular applications of ultrasound (Edited by R.S.
Reneman). pp. 266-282, North-Holland/American Elsevier, Amsterdam-
London-New York.

Hatle, L., Angelsen, B., and Tromsdal, A. (1979). Noninvasive assessment of atrioventricular pressure half-time by Doppler ultrasound.
Circulation 60, 1096-1104.

Hoeks, A.P.G., Reneman, R.S., and Peronneau, P.A. (1981). A multi-gate
pulsed Doppler system with serial data processing. IEEE Transac-
actions on Sonics and Ultrasonics SU-28, 242-247.

Hoeks, A.P.G., Reneman, R.S., Ruissen, C.J., and Smeets, F.A.M.
(1979). Possibilities and limitations of pulsed Doppler systems.
In Echocardiology (Edited by Ch.T. Lancée). pp. 413-419. Martinus
Nijhoff, The Hague-Boston-London.

Hoeks, A.P.G., Ruissen, C.J., and Reneman, R.S. (1980). A multi-gate
multi-purpose pulsed Doppler system. Fed. Proc. 39, 1177.

McLeod, F.D. (1967). A directional Doppler flowmeter. Digest of
Seventh International Conference on Medical and Biological Engineering, 213.

Mozersky, D.J., Hokanson, D.E., Sumner, D.S., and Strandness, D.E. Jr.
(1972). Ultrasonic visualization of the arterial lumen. Surgery 72,
253-259.

Olsen, C.F. (1977). Doppler ultrasound: a technique for obtaining
arterial wall motion parameters. IEEE Transaction on Sonics and
Ultrasonics SU-24, 354-358.

Peronneau, P.A., Bournat, J.P., Bugnon, A., Barbet, A., and Xhaard, M,
(1974). Theoretical and practical aspects of pulsed Doppler flowmetry: real-time application to the measure of instantaneous velocity profiles in vitro and in vivo. In Cardiovascular applications
of ultrasound (Edited by R.S. Reneman). pp. 66-84, North-Holland/
American Elsevier, Amsterdam-London-New York.

Peronneau, P.A., Hinglais, J., Pellet, M., and Léger, F. (1970).
Vélocimètre sanguin par effet Doppler à émission ultra-sonore
pulsée. L'Onde électrique 50, 369-384.

Pourcelot, L. (1977). Echo-Doppler Systems – Applications for the
detection of cardiovascular disorders. In Echocardiology with
Doppler applications and real time imaging (Edited by N. Bom). pp.
245-256, Martinus Nijhoff, The Hague.

Reid, J.M. and Baker, D.W. (1971). Physics and electronics of the
ultrasonic Doppler method. In Ultrasonographia medica (Edited by
J. Böck and K. Ossoinig). pp. 109-120, Verlag der Wiener Medizini-
schen Akademie, Wien.

Reid, J.M., Davis, D.L., Ricketts, H.J., and Spencer, M.P. (1974). A
new Doppler flowmeter system and its operations with catheter
mounted transducers. In Cardiovascular applications of ultrasound
(Edited by R.S. Reneman). pp. 183-192, North-Holland/American
Elsevier, Amsterdam-London-New York.

100

Reid, J.M. and Spencer, M.P. (1972). Ultrasonic Doppler technique
for imaging blood vessels. Science 176, 1235-1226.

Reneman, R.S. What measurements are necessary for adequate evaluation of the peripheral arterial circulation. Cardiovasc. Dis. in
press.

Reneman, R.S., Clarke, H.F., Simmons, N., and Spencer, M.P. (1973).
In vivo comparison of electromagnetic and Doppler flowmeters: with
special attention to the processing of the analogue Doppler flow
signal. Cardiovasc. Res. 7, 557-566.

Reneman, R.S. and Hoeks, A. (1977). Continuous wave and pulsed
Doppler flowmeters - a general introduction. In Echocardiology
with Doppler applications and real time imaging (Edited by N. Bom).
pp. 189-205, Martinus Nijhoff, The Hague.

Reneman, R.S., Hoeks, A., and Spencer, M.P. (1979). Doppler ultrasound in the evaluation of the peripheral arterial circulation.
Angiology 30, 526-538.

Reneman, R.S. and Spencer, M.P. (1974). Difficulties in processing
of an analogue Doppler flow signal: with special reference to
zero-crossing meters and quantification. In Cardiovascular applications of ultrasound (Edited by R.S. Reneman). pp. 32-42, North-
Holland/American Elsevier, Amsterdam-London-New York.

Reneman, R.S. and Spencer, M.P. (1979). Local Doppler Audio Spectra
in normal and stenosed carotid arteries in man. Ultrasound Med.
Biol. 5, 1-11.

Rice, S.O. (1954). In selected Papers on Noise and Stochastic Processes (Edited by W.Wax). pp. 133-294, Dover Publications - New York.

Sandmann, W., Peronneau, P.A., Schweins, G., Bournat, J., and
Hinglais, J. (1978). Turbulenzmessung mit dem Doppler-Ultraschall-
verfahren: Eine neue Methode der Qualitätskontrolle in der Arte-
rienchirurgie. In Ultraschall-Doppler-Diagnostik in der Angiologie
(Edited by A. Kriesmann and A. Bollinger). pp. 77-81, George Thieme
Verlag, Stuttgart.

Skidmore, R., and Follett, D.H. (1978). Maximum frequency follower
for the processing of ultrasonic Doppler shift signals. Ultrasound
Med. Biol. 4, 145-147.

Spencer, M.P. and Reid, J.M. (1979). Quantitation of carotid stenosis
with continuous wave (CW) Doppler ultrasound. Stroke 10, 326-330.

Spencer, M.P., Reid, J.M., Davis, D.L., and Paulson, P.S. (1974).
Cervical carotid imaging with a continuous-wave Doppler flowmeter.
Stroke 5, 145-154.

Strandness, D.E., Kennedy, J.W., Judge, T.P., and McLeod, F.D.
(1969). Transcutaneous directional flow detection: A preliminary
report. Am. Heart J. _78_, 65-74.

Wells, P.N.T. (1969). Physical principles of ultrasonic diagnosis.
Academic Press, London.

Wille, S.Ø. (1979). Numerical models of arterial blood flow. Thesis.
Institute of Informatics, University of Oslo.

CHAPTER 5

Continuous Wave Doppler Techniques in Cerebral Vascular Disturbances

L. Pourcelot

5.1 INTRODUCTION

The following report presents the synthesis of results obtained with the continuous wave Doppler examinations of the carotid and vertebral arteries since 1969. We have carried out 23.000 Doppler examinations of which 13.000 were of the carotid and vertebral circulations. More than 350 carotid occlusions, 850 carotid stenoses and up to 150 subclavian steal syndromes have been diagnosed.

5.2 DIFFICULTIES OF THE DOPPLER EXAMINATION

These result from several factors which are related to the operator, or the equipment, or else to the conditions of the examination itself.

The audible Doppler signal contains all the useful information and the ear can easily pick out anomalies such as the wide frequency spectrum of turbulent flow. However, it is only after long experience and after having recorded many velocity curves that one can interpret a Doppler signal with the ear without looking at its graphical tracing.

The probe should be correctly orientated (it should not be placed perpendicular to the vessel axis) and one should get used to locating in space the direction of the ultrasonic beam, and if possible the point of intersection between the beam and the artery. This requires some training but helps to avoid confusing two neighbouring vessels.

The identification of some arteries poses tricky problems: learning how to solve them helps to improve the quality of the examination.

This is a particularly important point to assess a complete occlusion especially at the level of the carotid bifurcation.

Although the detection of the Doppler signal is usually quite easy, several factors affect the quality of the recordings: interference between venous and arterial flow, orientation of the probe at a bad angle, movements of the probe on the patient, a defective probe, a badly tuned instrument, external noise, etc. One should be able to recognize those perturbations and eliminate them.

Some patients who are obese or thick-set or emotionally upset are difficult to examine. One can use simple tests in order to increase the diagnostic accuracy: local compressions, hyperemia tests, etc.
of the examination seem suspicious. It is always better to start another examination than to give a report based on unreliable data.

5.3   EXPLORED ARTERIES

All the superficial arteries are detected by the Doppler beam, provided that the reflected energy is sufficient and that the ultrasonic beam does not insonate any bone or air. In practice, all the arteries of the limbs can be detected. For the cerebral circulation, only the vessels of the neck can be presently studied. A complete examination includes the common, external and internal carotid arteries, the vertebral arteries, and finally the ophthalmic artery or its branches (nasal, supra-orbital). The subclavian artery also must be included in the systematic evaluation of cerebral vascular patients.

The carotid arteries are studied in all the parts which are accessible to the Doppler beam. In doubtful cases (thrombosis (= occlusion) of the internal carotid artery for example), the external carotid artery can be identified by using a simple test: we compress the temporal, occipital or facial branches of the external carotid artery. When the compression is released the flow increases immediately in the detected artery, only if the probe is located on the external carotid artery. The vertebral artery is studied at the level of the occipital loop. The Doppler probe is placed below the mastoid and directed towards the contralateral orbit. However, every time a stenosis of the vertebral ostium is suspected, an examination of the ostial area will be required so as to detect turbulence or increased velocity at that level. Note that the vertebral and internal carotid arteries can be studied through the mouth, after having sprayed a local anesthetic on the pharynx.

5.4   CEREBRAL CIRCULATION

5.4.1   Cerebral Circulatory Resistance

The velocity curves recorded at the level of the major superficial arteries, vary according to the areas studied. This variation results from two important causes: on the one hand the circulatory resistance does not have the same value and depends upon the tissues supplied, on the other hand the blood movement does not occur at the same time in all the parts of the vascular system.

In practice, one observes in the arteries supplying the brain a continuous pattern of flow upon which a systolic component is super-imposed. The circulatory resistance of the brain is sufficiently low for the diastolic pressure to ensure a large continuous flow in the internal, common carotid and vertebral arteries (Mol et al, 1971). The brain is therefore permanently supplied with a systolic increase of the flow, the continuous flow representing up to 60% of the total flow. The areas supplied by the external carotid artery and the ophthalmic artery branches have a much higher resistance than that of the brain. For this reason the diastolic component of the flow is normally close to zero in these arteries. The same can be said for the circulation of the upper limbs which gradually stops during diastole, so that in case of subclavian steal the flow in the homolateral vertebral artery will be negligible or nil during the diastolic phase, whereas the steady flow will remain large in the healthy vertebral artery which is in normal relation with the brain (Fig. 5.1).

5.4.2   Index Of Cerebral Circulatory Resistance

The instantaneous cerebral blood flow is the sum of two terms:

- a resistant flow P/R proportional to the instantaneous blood pressure P, that is to say consisting of systolic and diastolic flows (continuous flow)
- a compliant flow $C\frac{dP}{dt}$ linked to the quantity of blood stocked up by the increase in volume of the arterial system during each systole. This term is directly dependent upon arterial elasticity and is only linked to pressure variations in the arteries. It will be more noticeable in young subjects than in atherosclerotic subjects (Fig. 5.2). In the case of an increase in circulatory resistance, the term P/R decreases so that the continuous diastolic flow tends to disappear. Thus we see that by calculating the proportion of systolic flow in the total flow of the carotid system, an index of circulatory resistance can be evaluated. This calculation is preferably made on the velocity curve of the common carotid artery for two reasons:
- that artery is easily recorded
- the index is very useful in cases of stenosis or thrombosis of the internal carotid artery, in which cases the recording of the internal carotid artery flow is either impaired or impossible.

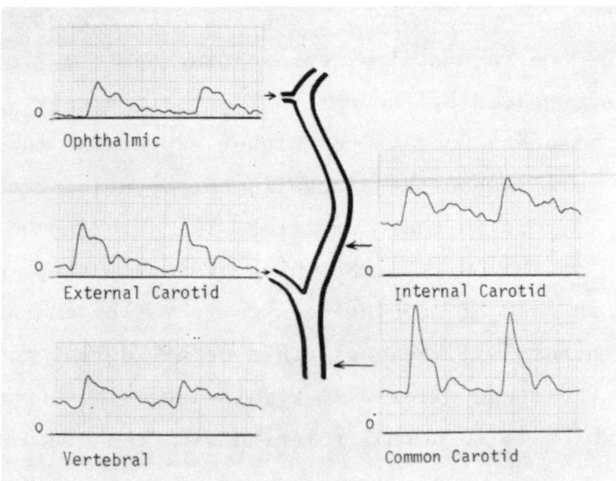

FIG. 5.1 Normal velocity curves in carotid, ophthalmic and vertebral arteries. Note the large continuous flow in arteries supplying the brain.

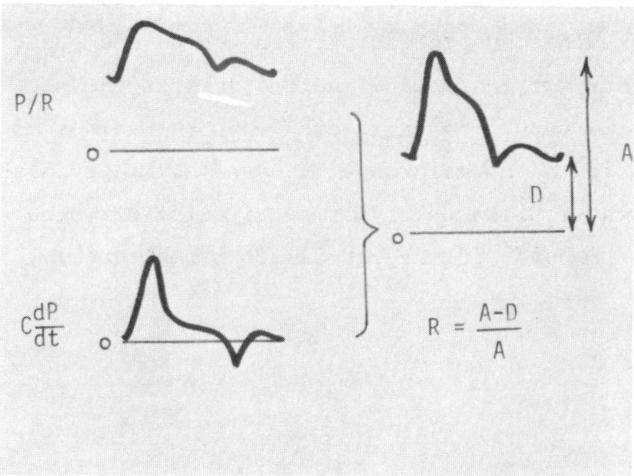

FIG. 5.2   The flow in the common carotid
artery is the sum of a resistant flow P/R
and a compliant flow $C\frac{dP}{dt}$

Fig. 5.3 shows the relationship between instantaneous flow and pressure in the carotid arteries. Pressure oscillates, during each cardiac cycle, between the diastolic value P_d and the systolic value P_s. In the internal carotid artery, the circulatory resistance is very low and we detect a large diastolic flow. On the other hand a high pressure threshold must be reached in the external carotid artery before ensuring a flow owing to the high circulatory resistance. The diastolic continuous flow is nil or negligible in that vessel. The resistance of the common carotid artery is halfway between these two values. If A is the maximum systolic flow and D the diastolic value, the resistance index which we have determined reads $R = \frac{A - D}{A}$. It would be advisable to calculate the ratio of the systolic and diastolic surfaces in order to keep to a minimum the influence of the term $C\frac{dP}{dt}$. That calculation being time-consuming we keep it exclusively for young patients.

This circulatory resistance index is easy to calculate, it depends little on the recording conditions, and it is determined separately for each side of the patient. In a normal subject, the indices should be perfectly symmetrical. Results are normal if there is no asymmetry and when values are between 0.55 and 0.75 for the amplitude ratio, and between 0.30 and 0.45 for the surface ratio. An increase in the index R corresponds to a slowing down in the cerebral flow, whereas a decrease makes one suspect an arterio-venous aneurysm, angioma, etc. An asymmetry of resistance indices between the patient's right and left side is very important, even when the values of those indices are within the normal range. Thus an asymmetry observed in velocity curves of high quality is a criterion of vascular disease (stenosis, thrombosis, intracerebral occlusion, etc.).

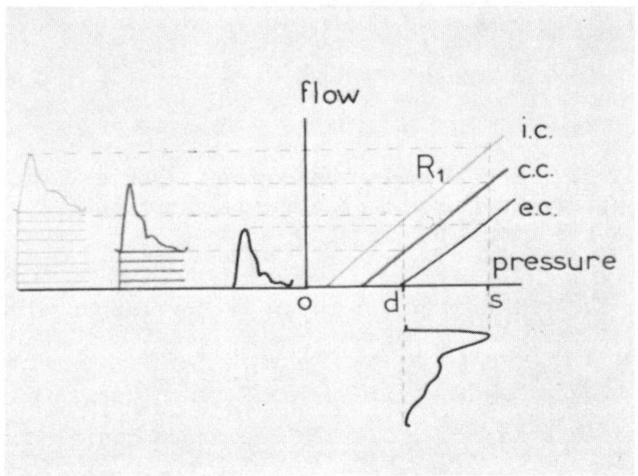

FIG. 5.3 Schematic relationship between instantaneous flow and pressure in carotid arteries. The low circulatory resistance (R_1) of the internal carotid artery explains the large diastolic flow in that artery.

5.4.3   Influence Of Diastolic Pressure

Whenever there are alterations in arterial pressure, regulation of the brain resistance comes into play to keep the cerebral blood flow at a constant value. This effect occurs if the pressure values remain between definite limits or if the pressure change is not too fast. For example, if one squats for 10 to 20 seconds and then moves quickly to a standing position, an immediate lowering in diastolic pressure is seen, giving rise to a transitory disappearance in the

diastolic cerebral flow (Fig. 5.4). This phenomenon is caused by the
reduction of the circulatory resistance of the lower limbs through
reactional hyperemia.

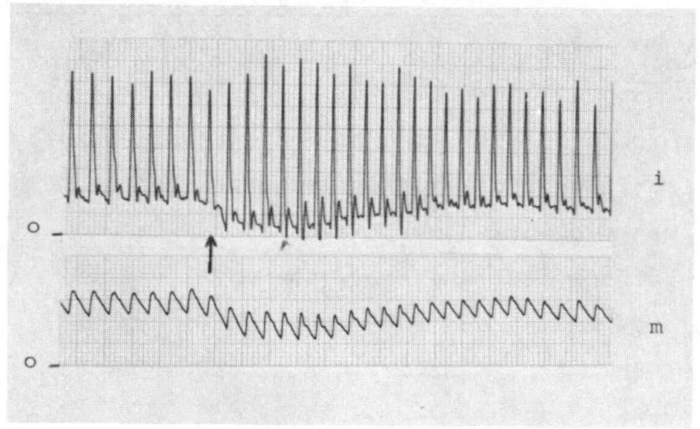

FIG. 5.4 Instantaneous (i) and average (m)
flows in the common carotid artery. The
arrow indicates the time at which the
subject stands up. Note the transitory
decrease in diastolic flow.

5.5   CAROTID CIRCULATORY DISEASE

5.5.1   Complete Occlusion Of The Common Carotid Artery

In these fairly rare cases, a Doppler examination shows the complete
disappearance of flow in the common and eventually in the internal
and external carotid arteries. When the occlusion only involves the
common carotid artery, flow is detected in the internal and external
arteries due to anastomotic pathways (thyroid artery for example).
The ophthalmic circulation can be reversed or will continue to flow
in its physiological direction (but reduced) according to the degree
of superficial anastomosis and the value of the pressure in the
carotid artery at the take-off of the ophthalmic artery.

5.5.2.   Stenosis Of The Common Carotid Artery

Diagnosis can be made on the basis of turbulent flow along the common
carotid artery. The most frequent locations are the ostium of the

artery and the bulb. The latter is easily accessible with ultrasound but in the case of the former difficulties result from interferences from neighbouring vessels.

5.5.3 Thrombosis (Complete Occlusion) Of The Internal Carotid Artery

In the case of complete occlusion of the internal carotid artery the diagnostic criteria will vary a little according to the different locations of the thrombosis, which are:
- the carotid bifurcation: these are the most frequent cases
- between the bifurcation and the take-off the ophthalmic artery
- at the end of the carotid artery distal to the origin of the ophthalmic artery.

5.5.3.1 Asymmetry Of Carotid Artery Flows

The criterion which is the same for all the different locations of internal carotid artery thrombosis is the asymmetry of the flow in the common carotid arteries (Pourcelot, 1976). The increase in circulatory resistance reduces the resistant component of the carotid flow, particularly the steady flow. The latter tends therefore to disappear which causes an increase in the resistance index R on the thrombosed side (Fig. 5.5). A recent study carried out on 270 cases of internal carotid artery thrombosis diagnosed in our department gave the following results:
- more than 50% of the indices are between 0.90 and 1
- most indices are between 0.66 and 1, with only 6 cases (about 2%) less than 0.75
- the average of R is 0.86

On the opposite side to the thrombosis we found 3 possibilities:
- thrombosis (case of bilateral thrombosis) in 8 cases (3%)
- carotid artery stenosis in 58 cases (21.5%)
- normal circulation in 204 cases (75.5%)

For those 204 contralateral carotid arteries without any lesion detected at the level of the carotid bifurcation, the values of R range

from 0.50 to 0.95 with an average value of 0.72. The value of R on
the supposedly healthy side is always smaller than the one on the
thrombosed side. The two distributions corresponding to the indices of
the thrombosed side and those of the healthy side are sharply differentiated, which confirms the advantage of such a measurement.

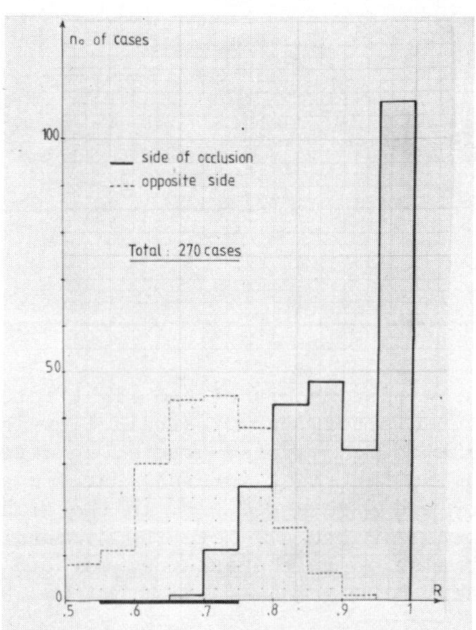

FIG. 5.5 Statistical assessment
of values of the index of circulatory resistance R in patients with carotid artery
thrombosis (270 cases). The dotted line represents the non-
thrombosed side, the continuous
line the occluded side.

5.5.3.2   Internal Carotid Artery Flow

The key for the diagnosis of occlusion (Planiol et al, 1972; Planiol
and Pourcelot, 1973) of the internal carotid artery at the level of
the bifurcation lies in assertion that the latter is not detected
(Fig. 5.6). It is important to be sure that the only detectable vessel
is the external carotid artery. Being able to distinguish between the
internal and external carotid arteries in all the patients examined

for cerebral vascular disease, requires some experience and long
practice.

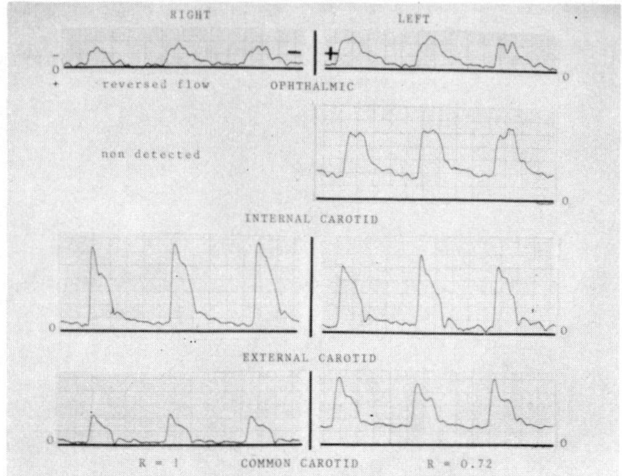

FIG. 5.6   Thrombosis of the right internal
carotid artery: no diastolic flow in the
right common carotid artery, no detected
flow in the right internal carotid artery and a reversed flow in the right
ophthalmic artery. Note the symmetrical
circulation in the external carotid
artery.

When the thrombosis lies several centimeters distal to the bifurcation
(Fig. 5.7) there is a slight compliant flow $C\frac{dP}{dt}$ into the proximal
part of the internal carotid artery. It consists of a positive peak
during systole (increase in the diameter of the vessel) followed by
a diastolic backflow (return to normal size) which is quite typical,
but is very unusual at that level. When the carotid artery thrombosis
occurs distal to the take-off of the ophthalmic artery, one can observe an internal carotid artery flow greater than the previous one
since it consists of:

- a compliant flow as mentioned above

- and the ophthalmic artery flow.

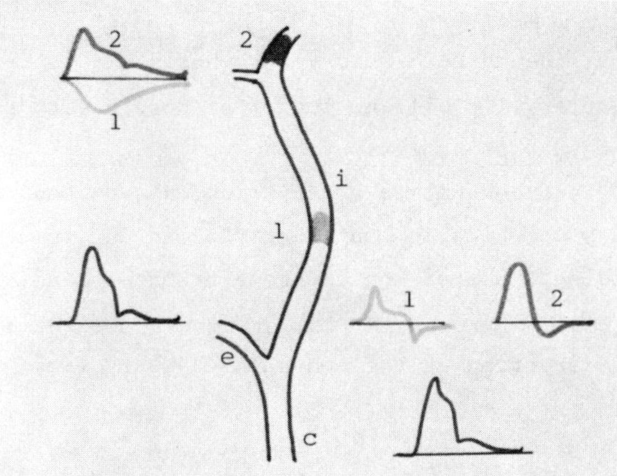

FIG. 5.7 Typical velocity curves in case of
total occlusion of the internal carotid
artery located several centimeters distal
to the bifurcation (1) and distal to the
origin of the ophthalmic artery (2). c, e
and i are the common, external and internal carotid arteries, respectively.

5.5.3.3   Ophthalmic Artery Flow

The direction of the ophthalmic artery flow is a noteworthy criterion
in the case of internal carotid artery occlusion (Muller, 1973). A
study of 214 cases of total occlusion of the internal carotid artery
gave the following results:

- reversed ophthalmic flow: 168 cases or 78%
- physiological direction: 37 cases or 17.8%
- undetected flow direction: 9 cases or 4.2%

Reversed flow depends on several factors:

- the development of anastomotic pathways between the homolateral or
  contralateral external carotid artery, the contralateral ophthalmic
  artery and the distal branches of the ophthalmic artery (Fig. 5.8)
- the existance of sufficient pressure in these collateral pathways:
  in the case of stenosis or thrombosis of the external carotid artery, the inversion of ophthalmic artery flow will be more difficult
  to obtain

- the value of the pressure in the distal part of the internal carotid artery: that pressure is high if the circle of Willis is functional
- the extent of the carotid artery thrombosis. If it reaches the ophthalmic artery, it will no longer be possible to get a physiological flow in that artery.

Each time reversed ophthalmic artery flow is detected, the superficial branches which could contribute to the anastomotic flows should be studied. Compression of these branches leads to a reduction or a disappearance of reversed flow or even sometimes a return to the physiological direction of the ophthalmic artery flow.

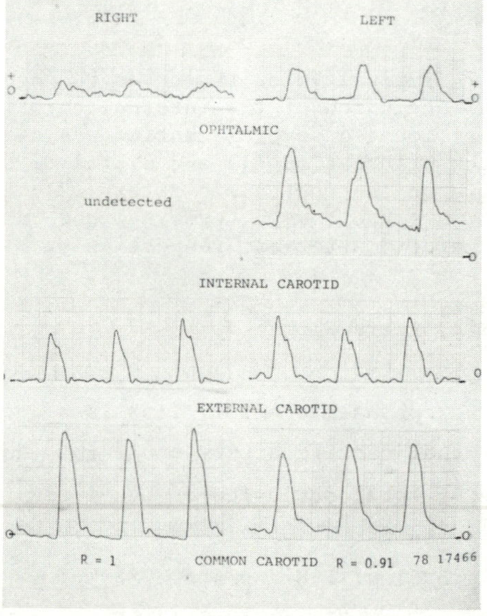

FIG. 5.8 Thrombosis of the right internal carotid artery with anterograde flow in the ophthalmic artery (physiological direction). No diastolic flow in the right common carotid artery. There is an abnormal increase in circulatory resistance on the left side.

5.5.3.4   Causes Of Failure

Diagnostic accuracy is quite high in the Doppler detection of carotid artery thrombosis, and is between 90% and 100% in experienced teams. The main causes of error are the following:

- direct detection of a big vertebral artery which is mistaken for the internal carotid artery, the probe being located on the carotid bifurcation. Such a mistake does not occur if one directs the probe systematically toward the inside of the neck when trying to detect the internal carotid artery

- the existence of continuous flow in the external carotid artery which simulates an internal carotid artery flow (hyperthermia, obese patients with erythrosis of the face, functional anastomotic pathways). A systematic compression of the external branches usually helps to dispel any doubt, if it impairs the flow in the detected artery. An asymmetry of the common carotid artery flow is always seen in such cases

- in some patients the internal carotid artery is not detected, while arteriography reveals a very tight stenosis. Here, the Doppler investigator is faced with two problems: the low density of energy backscattered at the level of the stenosis (greatly reduced blood volume available), and the possibility of marked attenuation of the ultrasonic wave due to the calcium within the plaque or the arterial walls. The problem is dealt with firstly by trying to explore the internal carotid artery distal to the stenotic area whenever this is possible. In the future the association of vascular imaging with the Doppler investigation will be helpful. This last cause of error is rather unusual but it is important to reduce it as much as possible, since some patients might be denied surgery on the basis of the Doppler investigation while their lesion is still operable.

5.5.4   Internal Carotid Artery Stenosis

5.5.4.1   General Comments

A reduction in vessel diameter does not systematically result in a disturbance of blood flow. First a narrowing of the vessel leads to an increase in flow velocity. The pressure drop remains moderate and the effect upon flow is almost nil. The stenosis is called non-functional.

For a greater reduction in the vessel diameter turbulent flow appears with a detected bruit during ausculation. Turbulence is associated with a loss of pressure which corresponds to an increase in the circulatory resistance: the stenosis is now functional.

5.5.4.2   Diagnosis Of Carotid Artery Stenosis

The diagnosis of a carotid artery stenosis is founded upon more numerous criteria than for occlusion (Fig. 5.9).

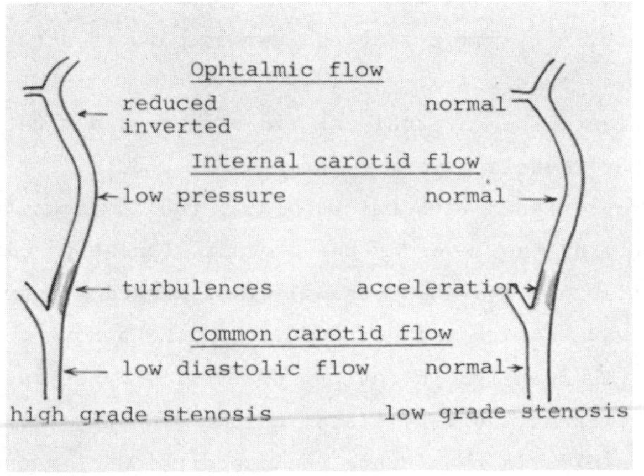

FIG. 5.9 Doppler criteria in high-grade and
low-grade stenosis of the internal carotid artery.

If the stenosis lies at the level of the bifurcation, one will detect it by moving the probe from upstream to downstream:
- in the common carotid artery there is a variable decrease in the steady flow. The resistance index R is therefore all the more asymmetrical as the stenosis is more functional. This index can for a long time keep a value situated in the normal range in the

course of the evolution of the stenosis
- at the stenosis or just distal to the stenosis:
  . an area of increased velocity which results in an increase of
    the Doppler frequency
  . an area of turbulence corresponding to a typical overlapping of
    two signals: a high-frequency signal due to the increased velo-
    city and a low-frequency sound corresponding to the vibration of
    the arterial walls
- distal to the stenosis:
  the velocity curves are damped if there is a tight stenosis. The
  area of turbulent or accelerated flow widens to a greater or lesser
  extent, according to the anatomical conditions. The velocity tracings are disturbed in the area close to the stenosis due to the
  flow instability. Spectrum analysis shows a characteristic broadening of the Doppler spectrum
- ophthalmic artery flow inversion is an important criterion, but it
  is found in less than 20% of carotid artery stenoses
If the stenosis is located far from the bifurcation, it can only be
detected by indirect signs such as the increase in the circulatory
resistance, and the disturbance in ophthalmic artery flow. It is
obvious that in this case it will not be possible to detect a non-
functional stenosis by the Doppler examination.

5.5.4.3   Accuracy Of The Doppler Investigation

In cases of carotid artery stenosis, the diagnostic accuracy is directly linked to the functional value of the stenosis. We can conclude as follows:
- a functional stenosis of more than 60% occurring at the bifurcation
  is detected with a degree of accuracy close to that of the occlusion; that is to say between 90% and 100%. The erroneous cases are
  those where the diagnosis of occlusion is made in cases with a
  very tight stenosis, or cases where the flow is not detected due to
  interference of calcium deposits
- the non-functional stenosis (reduction in diameter of less than 60%)
  can only be detected if it lies in an area accessible to ultrasound

in order to reveal a local acceleration of flow. Such stenoses are well detected for narrowing, ranging between 40% and 60% of the arterial lumen. Below 40% it seems difficult to get an accurate diagnosis by Doppler investigation since the resulting acceleration is difficult to differentiate from that due to a change in the angulation of the Doppler beam with regard to the vessel axis when the probe is moved along the artery.

- a functional stenosis located far from the bifurcation will be all the more easily detected as it will be more serious.

To conclude, we have seen that Doppler accuracy depends on two major factors:

- the functional degree of stenosis: diagnostic accuracy increases with that parameter
- the localization of the stenosis: a stenosis occurring at the bifurcation is more accurately detected than if it is far from it. This is an important result since the stenoses which are surgically accessible are better demonstrated by Doppler examination.

5.5.4.4 Causes Of Failure

There are quite a few causes of failure. Some of them have already been mentioned:

- localization of non-functional stenosis
- acoustical shadows due to calcium or atheromatous plaques: the association of real-time B-scan and Doppler investigation could be a solution to this problem
- the overestimation of the Doppler frequency in a healthy artery, with a change in angulation between the beam axis and the vessel axis
- the isolated stenosis of the external carotid artery: an unusual case in which one can conclude that there is an internal carotid artery stenosis if one does not use the compression test
- turbulent flow in the superior thyroid artery is a common trap which one must rule out before concluding that there is an internal carotid artery stenosis.

It is important to point out that the functional value of a stenosis

depends on the flow which passes through it. Consequently young
subjects in whom cerebral circulatory resistance is normal will
have for the same degree of stenosis, a more functional stenosis
than elderly subjects.

5.5.4.5 Follow Up Of Stenosis, Carotid Artery Surgery

The follow up of stenosis can be easily carried out by Doppler examination, but has obviously little interest since, whatever the
degree of stenosis, a stroke may result.

In operated patients, the Doppler follow up is easy and powerful
in the early post-operative phase to check the permeability of the
artery. Fig. 5.10 shows findings obtained before and after endarte-
riectomy of the internal carotid artery.

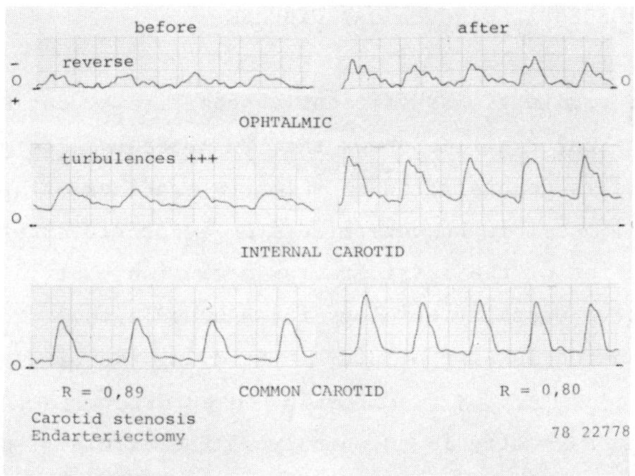

FIG. 5.10 Internal carotid artery stenosis
in a patient seen before and after surgery: turbulence at the level of the
stenosis, low pressure flow distal to
the stenosis (velocity curve), reversed
flow in the ophthalmic artery, index
R : 0.89. After surgery a Doppler examination is almost normal, except the resistance index which remains higher than
0.75

There is a marked improvement, with a decrease in circulatory resistance, disappearance of turbulence and return to a normal ophthalmic artery flow direction. On a long term basis, this survey can detect the reappearance of a new stenosis. Note that the Doppler examination is very helpful for the control of the patency of extra-intracranial anastomoses.

5.5.4.6 Conclusions

Following a properly carried out Doppler investigation, the diagnosis of carotid artery stenosis or thrombosis can be made with a high degree of reliability. However, it must be pointed out that a negative Doppler examination does not rule out the possibility of a nonfunctional stenosis, or of an ulcerated plaque.

5.6   INTRACEREBRAL CIRCULATORY DISEASE

Is the Doppler examination sufficiently sensitive to detect an intracerebral occlusion? Practice shows that in numerous case a hemodynamic disturbance is demonstrated on velocity tracings of high quality. It is in fact impossible to determine from a carotid artery recording the exact location of the intracerebral occlusion, but in such cases, one can suggest a vascular etiology for a neurological problem. The Doppler examination is also helpful in studying the circle of Willis. If the circle of Willis is functional, the carotid artery flows are symmetrical. An asymmetry is shown only if the circle of Willis is not functional, leading to an increase in circulatory resistance on the pathological side.

Diffuse atherosclerosis causes an increase in the cerebral circulatory resistance. Elderly patients often show a high resistance index without any asymmetry of flow in the carotid arteries. There is no sign of functional vascular disease at the level of the neck in such cases and classical results show a bilateral collapse of diastolic steady flow.

A decrease in cerebral circulatory resistance leads to an abnormal rise of the diastolic continuous flow in the common and internal carotid arteries. These results are seen in case of CO_2 inhalation, arterio-venous aneurysm, angioma, etc.

5.7 VERTEBRAL CIRCULATORY DISEASE

5.7.1 Vertebral Artery Stenosis Or Thrombosis (Complete Occlusion)

The study of the vertebral circulation is more difficult than that of the carotid circulation for several reasons:
- there is often an asymmetry in the vertebral artery diameters
- the best recording place is at some distance from the vertebral ostium
- the exploration of the vertebral ostium is not always possible, depending upon the location of the latter with regard to the subclavian artery
- the occipital artery (branch of the external carotid artery) can be taken for the vertebral artery. This mistake is avoided by compressing the occipital artery with a finger during the study.

The diagnosis of vertebral artery disease can be asserted on hearing turbulence distal to the vertebral ostium and the recording of velocity curves with low amplitude and modulation. This diagnosis is still difficult and the degree of accuracy needs to be improved in the future.

The diagnosis of vertebral artery occlusion is theoretically easier but it is often difficult to differentiate between occlusion and hypoplasia, although the functional value of the artery may hardly be different.

5.7.2 Subclavian Steal Syndrome

The subclavian steal syndrome corresponds to a reversal of flow in the vertebral artery, whether it is transitory or permanent during the cardiac cycle (Von Reutern and Pourcelot, 1978). In normal recording conditions of the velocity curves, it is not possible to determine with certainty the direction of the flow in the vertebral artery since we do not know what part of the atlas loop is reached by the ultrasonic beam. For that reason, the diagnosis of the vertebro-subclavian steal relies on direct and indirect criteria.

5.7.2.1   Indirect Criteria

These criteria are:
- there may be an asymmetry of brachial artery pressures. However, it can be very small or even absent in some cases
- the velocity curves are of low modulation in the subclavian artery on the side of the subclavian stenosis or thrombosis
- sometimes a systolic bruit is heard with the stethoscope and an increased velocity or turbulence is detected with the Doppler probe. These results help to assess the existence of subclavian stenosis.
- the diastolic flow is reduced or nil in the vertebral artery on the side of the subclavian occlusion
- there is a strong analogy between the velocity curves of the subclavian and vertebral arteries on the pathological side in the case of a permanent steal.

5.7.2.2   Direct Criteria

The diagnosis is confirmed by a positive compression test of the brachial artery: a pressure cuff is placed around the arm on the suspected side and inflated for about 20 seconds at a value greater than the systolic pressure. In the case of a permanent subclavian steal, when the compression is released, the reactive hyperemia of the arm causes an increase in the subclavian artery flow and consequently a rise in the homolateral vertebral artery flow. Whereas, recordings made with normal subjects show that the vertebral flow is steady during the test.

5.7.2.3   The Different Stages Of Subclavian Steal

5.7.2.3.1 *Permanent steal*

In the case of total occlusion, or high grade stenosis of the subclavian artery proximal to the vertebral ostium, the circulation in the vertebral artery is always reversed (Fig. 5.11). The resistance

of the upper limb is usually high so that the vertebral artery flow
returns to zero in diastole: the asymmetry of vertebral flows is
quite evident. The hyperemia test shows a transitory diastolic flow
in response to the sudden decrease in circulatory resistance of the
arm. The asymmetry of brachial artery pressure is usually obvious,
but it can be less than 2,7 kPa in some cases with large vertebral
arteries. In the case of occlusion no bruit or turbulence is detected
at the level of the subclavian artery.

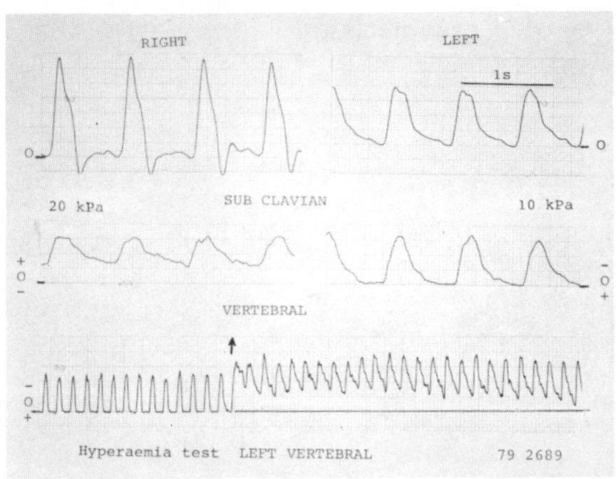

FIG. 5.11 Total occlusion of the left subclavian artery: no diastolic flow in the
  left vertebral artery with a tracing
  similar to that of the subclavian artery,
  asymmetry of brachial artery pressure.
  The hyperemia test is positive, the arrow
  indicates the time at which the compression is released (permanent steal).

5.7.2.3.2 *Transient subclavian steal*

In case of moderate stenosis of the subclavian artery, it is possible
to observe vertebral artery flow oscillating around the zero level,
with a reversed flow in systole and physiological anterograde flow
in diastole. During systole the stenosis is very functional owing to
the large amount of blood which flows toward the arm: the asymmetry
of pressure between the vertebral ostia is sufficient to give rise

to a reversed vertebral artery flow (Fig. 5.12). During the diastole
the flow falls to zero in the arm and the blood which passes through
the stenosis flows in the normal direction in the vertebral artery.
In such cases the average vertebral artery flow can be nil, which
explains why such a subclavian steal cannot be demonstrated by
angiography. During the hyperemia test, one observes that the transient subclavian steal becomes a permanent steal when the compression
is released. The asymmetry of brachial artery pressure can be low or
nil in such patients, but there is a subclavian bruit and turbulence
is detected by Doppler examination.

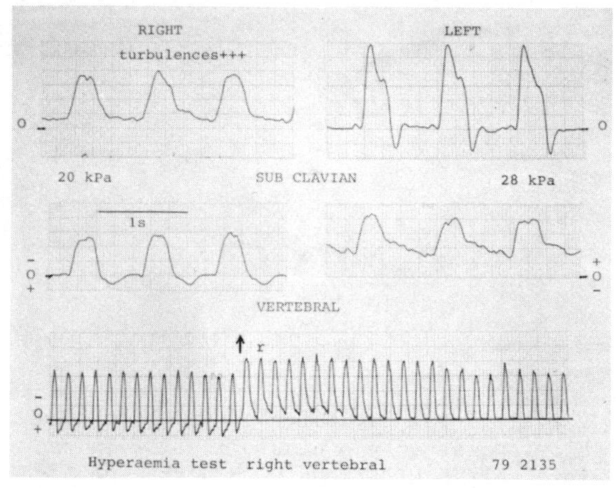

FIG. 5.12   Stenosis of the right subclavian
artery detected by turbulence, asymmetry
of brachial artery pressure and alternating flow in the right vertebral artery
(transient steal). During the hyperemia
test the transient steal becomes a permanent one.

5.7.2.3.3 *Latent subclavian steal*

In case of low grade stenosis of the subclavian artery detected by a
turbulence or increased velocity, the vertebral circulation permanently flows in the physiological direction, but a typical decrease
in the systolic peak of the vertebral artery velocity curve is seen

(Fig. 5.13). The hyperemia test brings about an inversion of the
systolic peak which creates a transient steal. The asymmetry of pressure is usually nil or very low at rest.

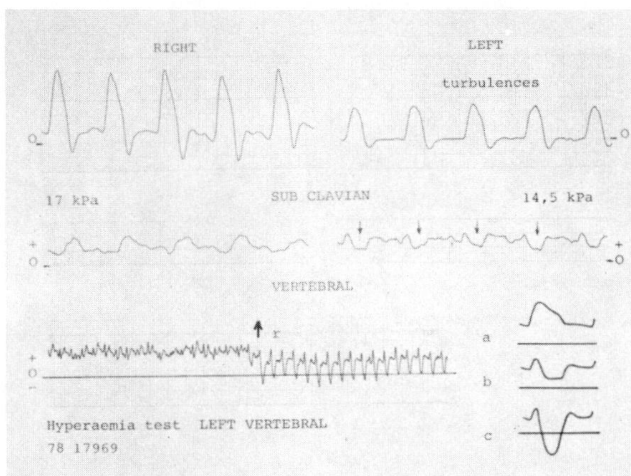

FIG. 5.13 Left subclavian artery stenosis
with a slight asymmetry of pressure,
turbulence and a reduction (b) of the
systolic peak in the vertebral artery
(arrows). The latent steal becomes a
transient steal during the hyperemia
test with an inversion of the systolic
flow (c).

5.7.2.3.4 *Subclavian stenosis without steal*

A low grade stenosis of the subclavian artery very slightly impairs
the vertebral artery flow at rest, and gives a reduction in the
systolic peak during the hyperemia test. There is no steal in that
case since the stenosis is almost non-functional (Fig. 5.14).

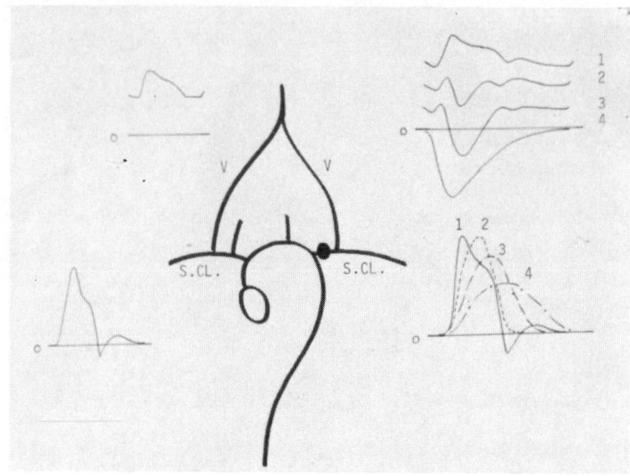

FIG. 5.14   Velocity curves for several
degrees of stenosis (1 to 4) of the left
subclavian artery proximal to the vertebral ostium. We see first a reduction
than an inversion of the systolic peak
in the vertebral artery.

To summarize we can say that there are several stages of disturbance of the vertebral artery flow in relation to the functional
value of the subclavian artery stenosis:

a) almost normal flow at rest   = no subclavian steal

b) reduction of the systolic peak = latent steal

c) inversion of the systolic peak = transient steal

d) total inversion of the flow   = permanent steal. The hyperemia
   test increases the functional value of the stenosis and allows
   us to pass from one stage to the next (Fig. 5.15).

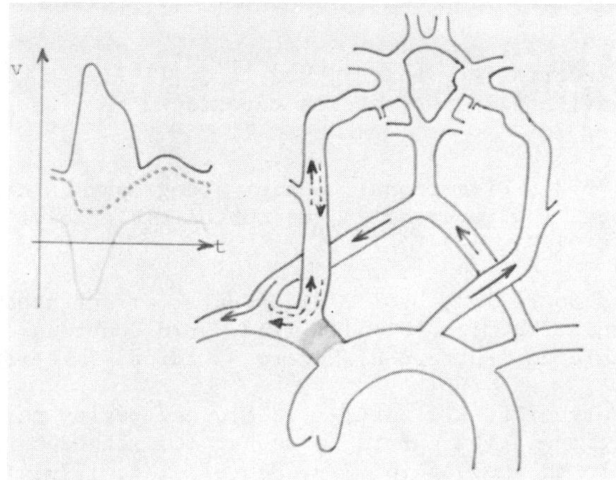

FIG. 5.15 In the case of total occlusion
of the innominate artery, the flow is
always reversed in the right vertebral
artery. In the carotid artery the systolic flow is either reduced or reversed
(transient steal).

5.7.2.3.5 Conclusions

Using the hyperemia test the accuracy of the diagnosis of subclavian
steal is 100% in our experience based upon 150 patients with 43
angiographies. Every patient seen for cerebral vascular disease is
suspected of having a subclavian steal syndrome. Indeed more than
70% of the patients are asymptomatic or show atypical clinical signs.
Using the procedure described above, 70 subclavian steal syndromes
were diagnosed in our department in 1978.

5.8   GENERAL CONCLUSIONS

Continuous wave Doppler ultrasonography is a non-invasive atraumatic
method of exploration of the peripheral vasculature which is accurate
and reliable when it is carried out by skilled examiners using equipment which is adapted to this purpose. Diagnostic accuracy is near
100% for carotid thrombosis and subclavian steal. In cases of carotid
artery stenosis, the area where surgery is possible is also the area
where the Doppler examination is most accurate.

REFERENCES

Mol, J.M.F., Frederix, L., and Rijcken, W.J. (1971). L'hématotachy-
graphie directionnelle des artères carotides. EEG, Neurophysiol.
Clin. 1, 331-337.

Muller, H.R. (1973). Directional Doppler sonography. A new technique
to demonstrate flow reversal in the ophthalmic artery. Neuro-
radiology. 5, 91-95.

Planiol, T. and Pourcelot, L. (1973). Doppler effect study of the
carotid circulation. pp. 104-111, IInd World Congress on Ultrasonics in Medicine, Rotterdam. Excerpta Medica, Amsterdam.

Planiol, T., Pourcelot, L., Pottier, J.M., and Degiovanni, E. (1972).
Etude de la circulation carotidienne par les méthodes ultra-
soniques et la thermographie. Rev. Neurol. 128, 127-141.

Pourcelot, L. (1976). Diagnostic ultrasound for cerebral vascular
disease. In Present and future of ultrasound (Edited by I. Donald
and S. Levi). pp. 141-147, Kooyker Scientific Publications,
Rotterdam.

Von Reutern, G.M. and Pourcelot, L. (1978). Cardiac cycle dependant
alternating flow in vertebral arteries with subclavian artery
stenoses. Stroke, 9, 229-236.

CHAPTER 6

The Clinical Use of Doppler Hematographic Investigation in Cerebral Circulation Disturbances

J. M. F. Mol

6.1   INTRODUCTION

In the present chapter I will report on the practical use of continuous wave (CW) Doppler blood flow velocity meters and of clinical
Doppler hematotachographic investigation in cerebral circulation disturbances in everyday practice. The expression hematotachogram (HTG)
is used to stress the fact that it is the velocity of the blood which
is registered rather than the volume flow per unit of time. We started this kind of investigation in 1969 with the first directional
apparatus built by Delalande and designed by Pourcelot.

The HTG is recorded in the Department of Clinical Neurophysiology
where the results are correlated with the case history, the neurological findings, the electroencephalogram and the echo examination
of the brain. In the course of the years the methods and the interpretation have become so reliable that they greatly influence the
decisions made in everyday clinical practice.

6.2   METHODS

6.2.1   Equipment

From the start it was obvious that listening to the Doppler sound
alone did not always give sufficient information to make a certain
diagnosis, especially in the carotid region. Therefore, from the
beginning the instantaneous mean velocity tracings obtained with the
Doppler device were registered on a physiological recorder.

130

By measuring liquid flow in plastic tubes, it soon became evident
that the velocity reading is only reliable if the velocity sensed by
the ultrasonic beam is in one direction. The simultaneous recording
of velocities in two directions can lead to variations in the instantaneous mean velocity tracing. This is especially important since
we use waveform analysis (see Section 6.2.3) to evaluate the cerebrovascular circulation. Therefore, we record two extra curves:

Curve B: the analog signal representing the velocity with which
       blood is moving away from the transducer crystals

Curve A: the analog signal representing the velocity with which
       blood is moving towards the transducer crystals.

Curve B-A, representing the difference between curve B and A, is the
normal output of the Doppler device as presented in most publications.
This B-A curve is only reliable if either the A or the B signal is
zero; i.e. if flow is in only one direction within the ultrasonic
beam (Fig. 6.1).

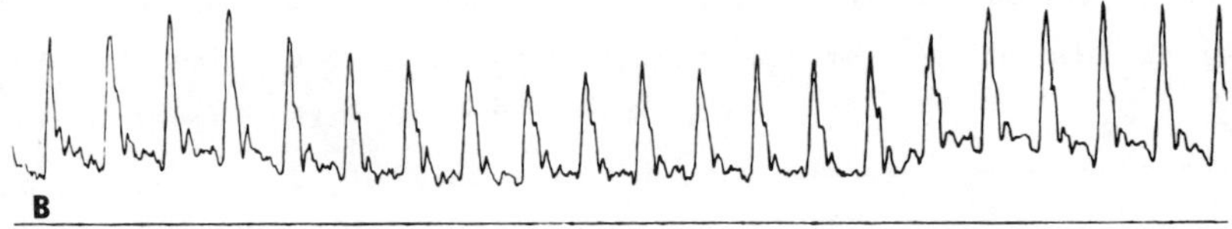

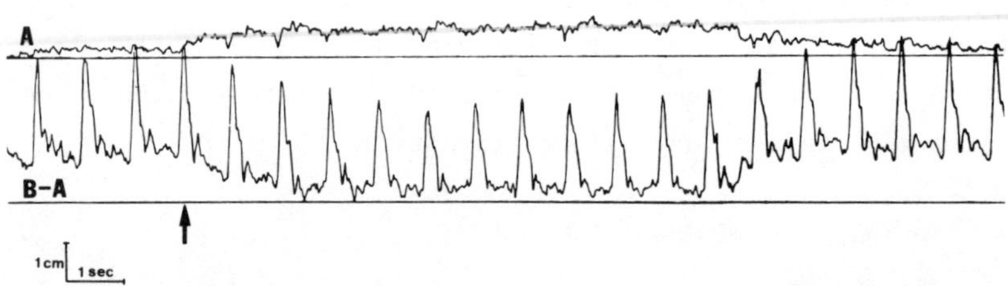

FIG. 6.1   HTG of a common carotid artery. Curve B-A is the
normal output of the Doppler device. This curve is only
reliable if either the A or the B signal is zero, i.e.
flow is in only one direction within the ultrasonic beam.
This is the case on the left and right hand side of the
curve. In case of venous contamination (arrow), which
can be seen in curve A, curve B-A decreases.

In the past years CW Doppler instruments (e.g. Versaton of Medasonics) have become commercially available which allow clear separation of two different flow directions. The B and A channels can be recorded separately and simultaneously, although this option is not always directly available in the commercial versions.

6.2.2   Calibration

To facilitate the comparison of curves from various persons which are recorded with different pieces of equipment, every recording is started with a calibration procedure. At a transmitting frequency of 4 MHz the amplification is adjusted so that a calibration signal, simulating a Doppler shift of 500 Hz, is associated with a deflection of 1 cm on the recorder. At a transmitting frequency of 5 MHz, a signal simulating a Doppler shift of 625 Hz produces this same deflection and with a frequency of 10 MHz, 1250 Hz will give the same pen deflection for the same blood flow velocity.

The calibration signal does not calibrate the Doppler apparatus as a whole because the signal is not entered in the first stage of the Doppler device. In order to calibrate the first stage of the apparatus, and in order to compare the performance of different hematotachographs, we also used another form of calibration. The calibration device has a wheel rotating exactly once per second. On the wheel a peg is fixed at a certain distance from the axle. On this peg a string of about 50 cm in length is fixed and the Doppler transducer is connected to this string. When the wheel is turned by a synchronous electromotor, the transducer, hanging in a glass of water, goes up and down in simple harmonic motion (Fig. 6.2A). The output of the Doppler apparatus will be a sinusoid in the B-A channel and an alternating half-sinusoid in the A and B channel. If the distance of the peg from the axis is 15 mm and the gain control of the Doppler device is set at standard amplification, the pen deflection must be 10 mm in the B and the A channel and 20 mm in the B-A channel. This calibration procedure is independent of the transmitter frequency of the Doppler apparatus (Fig. 6.2B).

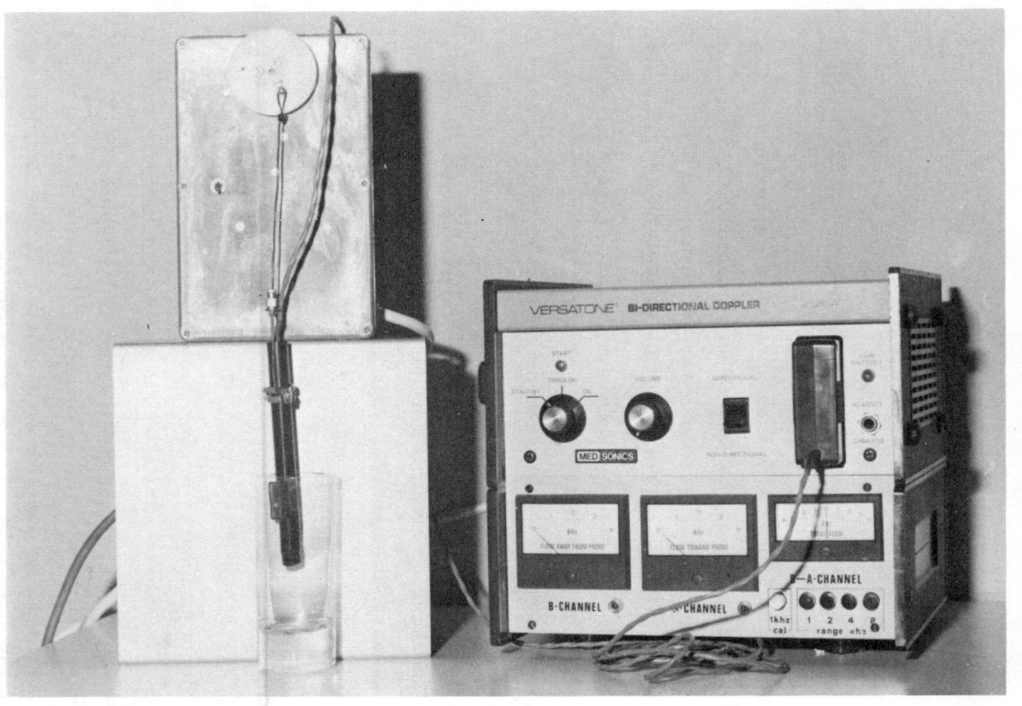

FIG. 6.2A   Calibration device (for explanation see text).

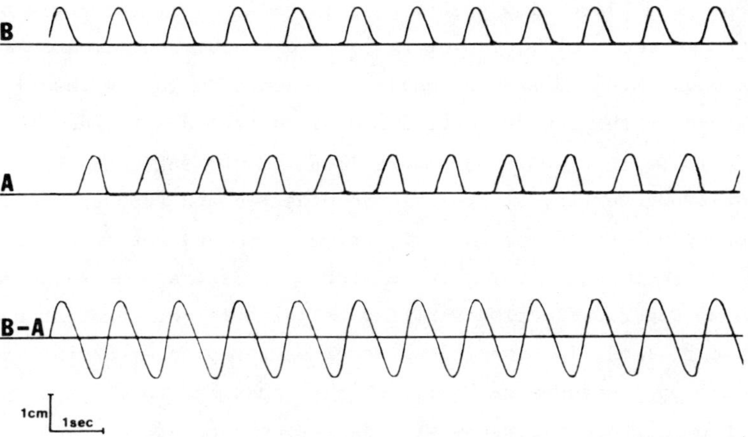

FIG. 6.2B   The output of the Doppler apparatus when the
probe is moved up and down in simple harmonic motion
in the device shown in Fig. 6.2A.

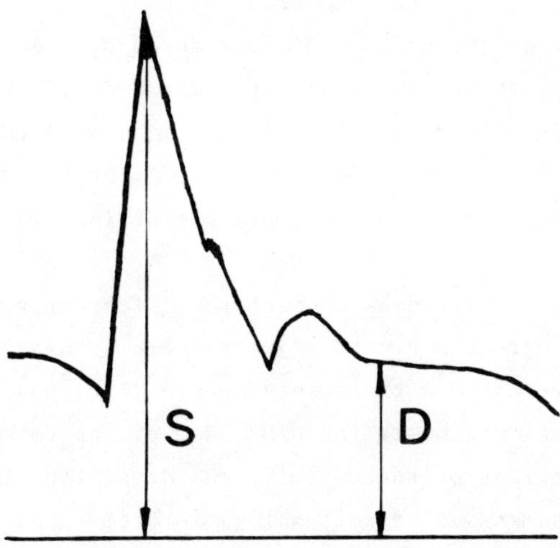

FIG. 6.3   Hematotachogram (HTG) parameters of the common carotid artery.
The mean value of the amplitude during diastole (D) is one half to one
third of the peak level during
systole (S).

With standard amplification a pen deflection of 1 cm means an instantaneous mean blood flow velocity of 11 cm sec^{-1} if the angle between the transducer and the vessel is 30°. To be able to compare examinations made with a different angle of insonation to the vessel the magnitude of the pen deflection can be corrected by means of tables to the value that would have been found if the angle had been 30°. In our laboratories, all values are measured and reported in the corrected values of the pen deflection in mm and not in blood flow velocity in cm sec^{-1}. The reason for this is that the relation between mm paper deflection and flow velocity is not constant under different blood flow conditions, such as turbulence. Having made our calibration procedure very rigid, it is possible to compare curves made from different persons with different apparatus.

6.2.3   Pulsatility Standardization And Adjustment For Age Variation

It was obvious as long as 9 years ago that the blood flow velocity in some vessels decreases in older people. We made HTG registrations of the carotid, vertebral and brachial arteries on both sides in 140 healthy individuals, uniformly distributed over 14 age-groups, each covering a 5 year interval. From the carotid HTG the maximum amplitude during systole (S level), and the mean amplitude during diastole (D level) have been determined (Fig. 6.3). From the vertebral and brachial arteries the maximum amplitude during systole was measured (S level).

The S and D amplitudes in mm of the different arteries were averaged per age-group and plotted with their standard deviations. These amplitudes decreased with increasing age (Fig. 6.4).

For clinical diagnosis the S/D index of the common carotid artery, being the quotient of the systolic and diastolic amplitudes of the instantaneous mean velocity tracings is used rather than absolute amplitude values. The S/D index in normal common carotid arteries is generally $\frac{2.2-3}{1}$.

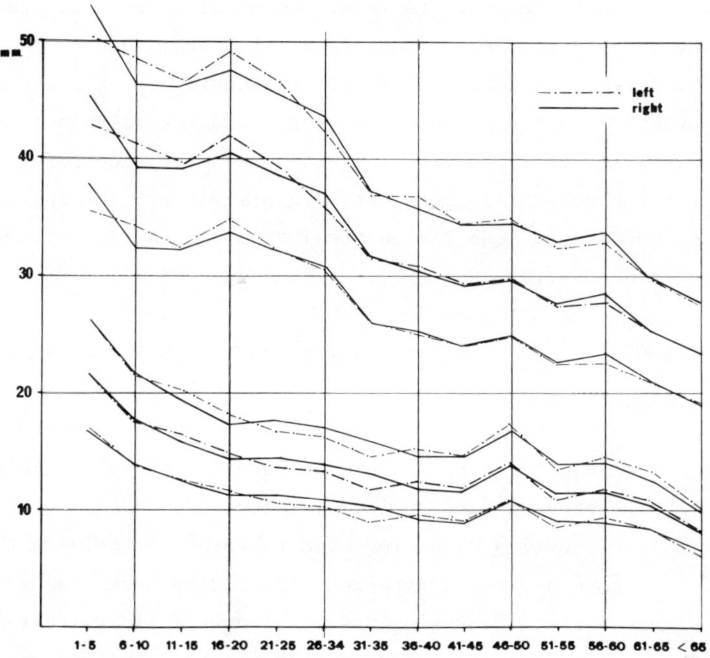

FIG. 6.4   Amplitude-age diagram of the S and D level
of the left and right common carotid artery, and
standard deviations determined in 140 normal subjects.

6.2.4   Routine Techniques

In a routine investigation of patients, bilateral recordings are
made of the common carotid artery and in most cases also of the internal and external carotid, the vertebral, the ophthalmic and the brachial arteries. Generally we first start recording the blood flow
velocity in the common carotid artery with the transducer positioned
low and high in the neck. Thereafter the blood flow velocity in the
internal and external carotid arteries is recorded separately. The
HTG of the vertebral artery is recorded just under the mastoid process, directing the ultrasonic beam towards the upper ridge

of the transverse process of the atlas. The HTG of the ophthalmic
artery is recorded transorbitally. The plethysmogram of one or both
ears and the electrocardiogram are recorded as well. The remaining
8 channels are often used to register the electro-encephalogram (EEG).

Routine recording of the brachial arteries was found te be necessary to differentiate between bilateral stenosis low in the common
carotid arteries on the one hand and heart and general blood vessel
disease on the other. Moreover, the brachial artery recording is important for the recognition of aortic insufficiency, low peripheral
blood flow resistance, subclavian steal syndrome and stenotic lesions
in the innominate artery.

6.3 STENOTIC LESIONS OF THE CAROTID AND VERTEBRAL ARTERIES

6.3.1 Introduction

The form of the HTG curve of the common carotid artery is characte-
rized by a high steep S wave during systole and a persistant flat
wave, or D level, during diastole. When the S and especially the D
level of both common carotid arteries differed by more than 25%,
then as a rule a stenotic lesion was found on the side with the lowest amplitudes, unless a space-occupying intracranial abnormality
was present.

6.3.2 Internal Carotid Artery

When the D level was decreased relative to the S level, the stenotic
lesion was usually localized in the internal carotid artery. When
the D level decreased slowly to zero or was zero during a period of
time, it could be assumed that the internal carotid artery was totally obstructed. Because the blood flow in the external carotid artery
is nearly zero during diastole, under normal circumstances the amplitude of the HTG of the common carotid artery during diastole is a
measure of the flow in the internal carotid artery where the blood
continues to flow between cardiac contractions. This diastolic amplitude has to be related to the normal value for the age group.

When the internal carotid artery is narrowed significantly or is totally obstructed, the HTG of the ipsilateral ophthalmic artery is abnormal. One usually finds a very low amplitude or a reversal of the flow direction that is to say, from the outside to the inside of the skull. However, when one finds an HTG of the common carotid artery that points in the direction of a stenosis of the internal carotid artery, but with an ophthalmic curve with a normal direction and even a high amplitude, then the chance is great that the stenosis is localized beyond the branching of the ophthalmic artery. This is especially so if no locally increased blood flow velocity can be found near the bifurcation or in the internal carotid artery, which can be expected to be present in case of stenosis in these areas. A localization of preference for lesions distal to the origin of the ophthalmic artery is in the distal part of the carotid siphon.

6.3.3   Common Carotid Artery

In the case of stenosis of the common carotid artery, the amplitude of the S and D levels on the affected side are more equally lowered, the steepness of the S wave is too low, while, as a rule, the flow direction in the ophthalmic artery is not reversed. The amplitude is - as can be imagined - mostly lower on the diseased than on the normal side.

6.3.4   Innominate Artery

When one finds a combination of the symptoms of a stenosis in the common carotid artery on the right side together with a pathologically lowered HTG of the right brachial artery with a pathological form, the existence of a considerable stenotic lesion of the innominate artery can be assumed. The differentiation of an innominate artery stenosis from a stenosis of the common carotid artery in combination with a subclavian artery stenosis, is discussed below.

6.3.5   Bilateral Internal Carotid Artery

When the D level of the common carotid HTG is lowered, both on the

138

 left and on the right side - in respect to the normal level for the
age group - and the brachial HTG is normal, one can expect a bilateral stenosis in the internal carotid artery, at least when the intracranial pressure is normal and neither the arterial blood pressure is
higher than 220/120 nor the hematocrit is seriously raised. Recording over or just distal to this stenosis shows a pronounced increase
in the amplitude of the HTG compared to a place more distal to the
site of obstruction (Fig. 6.5)

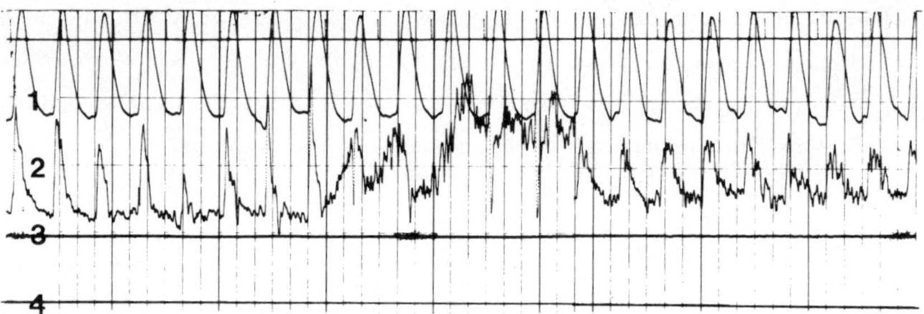

    FIG. 6.5   B-A curve of the HTG of a common carotid artery in a
    patient with a stenosis of the internal carotid artery
    (left). The D level is too low. When moving the probe up-
    wards high velocities are recorded over the stenosis (ar-
    row). The curve on the right hand side is the HTG of the
    internal carotid artery.

6.3.6   Vertebral Artery

The statistical findings regarding the vertebral artery are also used
as a reference to decide whether a measured amplitude in a patient is
abnormally high. This often happens, when the vertebral artery has a
compensatory function, in cases of stenosis of one or both carotid
arteries, or when it feeds arterio-venous aneurysms.

6.4   OBSTRUCTIONS OF THE INNOMINATE AND SUBCLAVIAN ARTERIES
6.4.1   Introduction

In the diagnosis of these lesions we make use of the so-called brachial reactive hyperemia procedure. It consists of inducing and

stopping reactive hyperemia of the underarm and the hand. This is realized by compressing the brachial artery, for instance, with a blood pressure cuff, and asking the patient to exercise his hand. After releasing the cuff the brachial blood flow increases considerably, especially during diastole. By inflating the cuff again the hyperemia is stopped.

6.4.2 Stenosis Of The Innominate Artery

The HTG of the brachial artery is pathological. The systolic amplitude is lowered. The rise time of the S wave is too long. The post-systolic reversed flow has disappeared. Mostly a considerable diastolic flow exists. The HTG of the common carotid artery is too low, especially during the systolic period. In carrying out the described procedure the amplitude of the HTG of the common carotid artery decreases chiefly during systole by deflating the cuff. It increases considerably by occluding the brachial artery (Fig. 6.6).

6.4.3 Total Occlusion Of The Innominate Artery

The brachial HTG is similar to that described above, but the changes are more pronounced. In the carotid and vertebral arteries the flow direction can be reversed. The amplitude of the carotid HTG decreases by inflating the cuff around the arm, demonstrating that the flow direction really was reversed. By carrying out the reactive hyperemia procedure the findings in the vertebral artery may be the same as in the carotid artery. By deflating the cuff the flow velocity increases in both the carotid and the vertebral artery, proving that the flow direction of the vertebral artery was also reversed.

In some cases only the vertebral artery shows a reversed flow which is distributed to both the brachial and the common carotid artery. Therefore, a normal flow direction in the common carotid artery is recorded with a low velocity. This is mainly the case if the circle of Willis is not functioning.

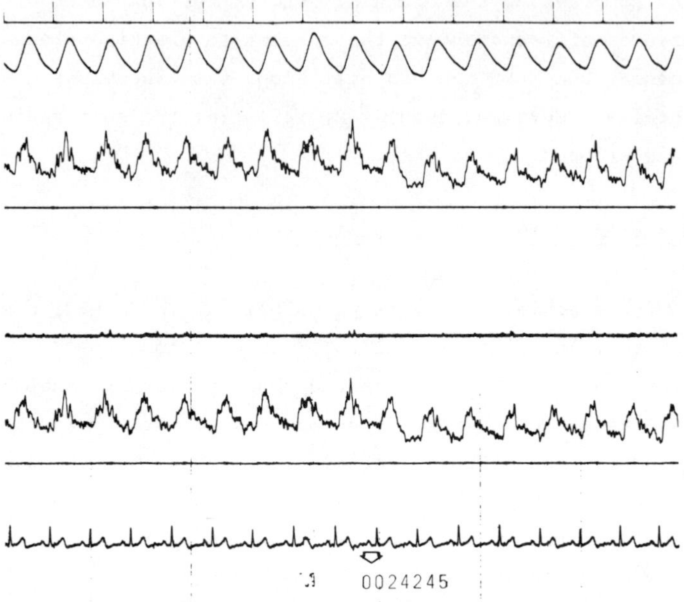

FIG. 6.6   HTG of the common carotid artery in a patient with a stenosis of the innominate artery. On
the left hand side of the recording the right brachial artery is compressed with a cuff. When it is
deflated (arrow) the amplitude of the HTG decreases
to the level that existed before the compression.

6.4.4   Stenosis Of The Subclavian Artery Proximal To The Branching Of The Vertebral Artery

In case of a subclavian artery stenosis proximal to the branching of
the vertebral artery the amplitude of the brachial HTG on that side
is decreased and the slope of the S wave is less steep than normal
and mostly a diastolic blood flow component is seen. The direction
of the ipsilateral vertebral flow is found to be reversed, flowing
downwards. The amplitude of the vertebral HTG increases - especially
the D level - when the affected arm is exercised and a reactive hyperemia is induced, according to the procedure described in section
6.4.1, showing that the arm is supplied with blood from the vertebral

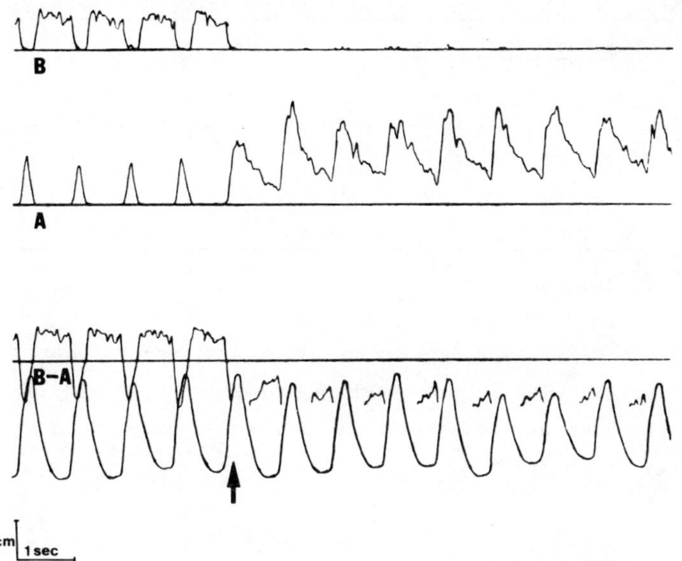

FIG. 6.7   HTG of a vertebral artery in a patient with
            stenosis of the subclavian artery proximal to the
            branching of the vertebral artery and a "basculating"
            steal. The ipsilateral brachial artery is compres-
            sed. Curve B shows that the blood flow has a normal
            direction during systole. Curve A shows a retro-
            grade flow direction in the systolic period. Upon
            deflating the cuff (arrow) the flow velocity in the
            vertebral artery increases in the retrograde direc-
            tion.

artery. This demonstrates that the vertebral flow is reversed and

these findings are typical for a subclavian steal syndrome. If the

same procedure is followed and the HTG of the other vertebral and

carotid arteries is recorded, an indication can be obtained con-

cerning the contribution of each of these vessels to the blood supply

to the steal. If the amplitude of the vertebral HTG increases during

brachial artery compression, the case can be described as a relative

steal. If by deflating the cuff the flow direction reverses we call

this a "basculating" steal. This can only be shown by angiography if

the contrast medium is injected after reactive hyperemia in the arm

is induced (Fig. 6.7).

6.4.5 Stenosis Of The Subclavian Artery Distal To The Branching Of The Vertebral Artery

The procedure as described in section 6.4.1 has no effect on the vertebral artery flow velocity.

6.4.6 Stenosis Of The Proximal Subclavian Artery And The Common Carotid Artery

The procedure may show a subclavian steal syndrome but hardly affects the pathological flow in the common carotid artery. This is in contrast to the combination of subclavian steal syndrome and an innominate artery stenosis.

6.5 TECHNIQUES OF VASCULAR CONFIRMATION

6.5.1 Introduction

In the course of the years, several procedures and tests have become common knowledge in our department. They are very helpful in proving that what is assumed, is in fact the case in differentiating between several possibilities and in answering questions that cannot be solved by the routine recordings. Our most important procedures and tests are the following.

6.5.2 Vertebral Artery Identification

If there is uncertainty that a supposed vertebral artery registration as recorded at the base of the skull is really a vertebral HTG, the common carotid artery on the same side is compressed during two or three heart beats (with plethysmogram registration of the homo-lateral ear). If during carotid compression, low in the common carotid artery, the amplitude of the plethysmogram disappears and the supposed vertebral HTG remains the same, one can be sure that it really is a vertebral HTG registration.

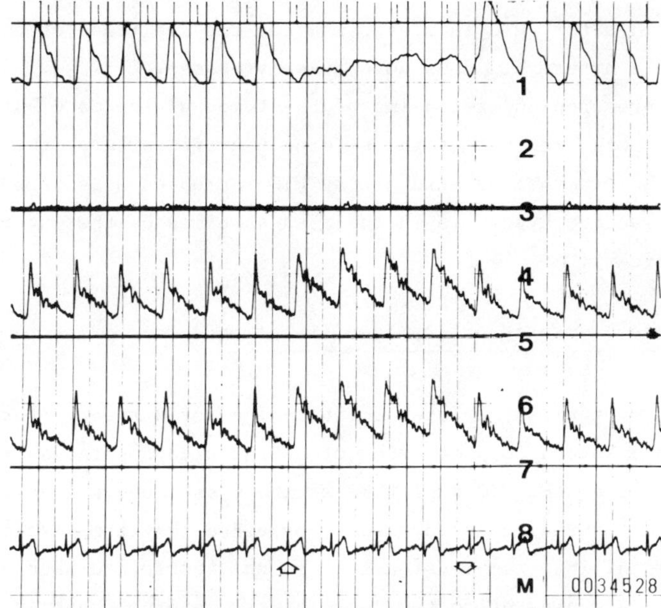

FIG. 6.8 HTG of a vertebral artery with an unusual
form. To demonstrate that the vessel is really
the vertebral artery, the common carotid artery
on the ipsilateral side is compressed (between
arrows). The amplitude of the plethysmogram of
the ipsilateral ear nearly disappears. The amplitude of the HTG, however, increases. This demonstrates that the vessel is the vertebral artery
and that there is an intracranial connection
between the vertebral and carotid arteries.

6.5.3 Intracranial Vertebral-Carotid Artery Connections

If the amplitude of the vertebral HTG increases during common carotid
artery compression, a connection between the carotid and the vertebral supply through the circle of Willis has been demonstrated.

6.5.4 External Carotid Artery Identification

If there is any doubt regarding the nature of the supposed external
carotid artery under examination, due to a relatively high diastolic
flow velocity for instance, two possibilities for identification are

at hand:

- if the patient swallows several times, the amplitude of the external carotid artery HTG increases whereas the amplitude of the internal carotid artery HTG does not markedly change.
- after compression and releasing of the temporal and mandibular
  arteries, both branches of the external carotid artery, the amplitude of the external carotid artery HTG increases as a sign of
  induced reactive hyperemia in the area supplied by the external
  carotid artery.

6.5.5 Internal Carotid Artery Identification

Compression of the contra-lateral internal carotid artery (for instance with a second Doppler probe of a hand-held apparatus) results
in an increase of the HTG amplitude of the vessel under investigation
if it is the internal carotid artery, provided that the circle of
Willis is open. This compression test is only performed when no
obstructions are present in the homo-lateral internal carotid artery.
In this condition compression of the contra-lateral internal carotid
artery has never lead to cerebral disturbances.

If the vessel under examination is the internal carotid artery
with a low diastolic amplitude, for instance when the vessel is narrowed in a place that is not accessable for the Doppler probe, compression of the vessel itself will lower the HTG amplitude of the
ipsi-lateral ophthalmic HTG artery with normal flow direction. If in
this situation the flow direction of the ophthalmic artery HTG increases, the supposed internal carotid artery was apparently the
external carotid artery.

6.5.6 Brachial Artery Identification

Occasionally there are patients with a double brachial artery. One
of them is the real brachial artery that feeds the forearm and the
hand. The other is a smaller branch for local blood supply. The flow
velocity of this latter one is much lower than that of the normal
brachial artery. If a moderate reactive hyperemia is induced by

letting the patient make a fist for several seconds, the real brachial artery shows a diastolic flow after releasing the fist, whereas
the HTG of the smaller branch does not alter.

6.5.7 <u>"False Flow Direction Indication" In The Ophthalmic Artery</u>

Not only inexperienced examiners may record a pathological flow direction in the ophthalmic artery if it is not present. The reason is
that the orbital walls are a good reflecting surface for the ultrasonic beam. It, therefore, can hit the ophthalmic artery from behind.
The HTG curve shows, in this case, flow in the wrong direction. To
be sure that one is recording the ophthalmic artery rather than the
concomitant vein slight pressure is exerted on the eyeball with the
probe. If the amplitude of the HTG curve does not diminish during
this procedure, it is likely that one records the artery.

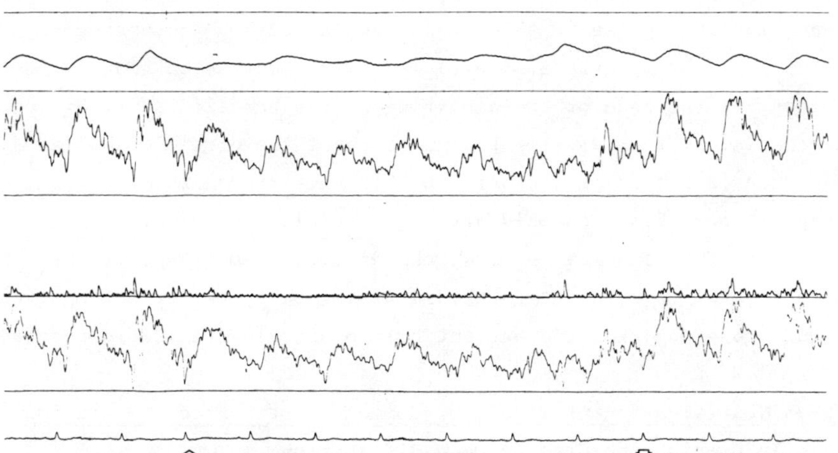

FIG. 6.9 HTG of an ophthalmic artery. The flow direction is
    reversed (curve B: away from the transducer). Compression
    of the branches of the external carotid artery on the ipsi-
    lateral side (between arrows) results in a decrease of
    the amplitude of both the plethysmogram and the HTG.

146

By compressing the branches of the external carotid artery in the
face, a discrimination can be made between a correct and a false
registration of the flow direction in the ophthalmic artery. By
compressing the branches of the external carotid artery the HTG
amplitude of the ophthalmic artery increases if the real flow direction is physiological - irrespective of the recorded flow direction.
If by compressing the external branches, the amplitude of the
ophthalmic HTG with a supposed normal flow direction decreases and
with a supposed pathological flow direction increases, a false
registration is demonstrated (Fig. 6.9).

6.5.8 Patency Of The Circle Of Willis

In section 6.3 it was described how a communication between the
vertebral and the carotid artery blood supply can be demonstrated.
To prove a good connection between the left and the right carotid
artery supply we compress the common, or the internal carotid artery
on one side during two or three heart beats (with a plethysmogram
recording of the ipsi-lateral ear). If this compression results in a
considerable increase of the amplitude of the carotid artery HTG on
the other side, the left-right connection of the circle of Willis
can be considered as patent (Fig. 6.10). This procedure of course,
is superfluous if one carotid artery is totally occluded, the vertebral artery HTG is normal or low, the diastolic amplitude of the
contra-lateral carotid artery HTG is high (nearly twice the amplitude
for the age-group) and the patient has no neurological deficiences.

6.5.9 Identification Of Total Low Common Carotid Artery Occlusion
In Combination With An Open Bifurcation

No flow is found in the common carotid artery but two flows can be
demonstrated at the usual places of the internal and external carotid
arteries but in opposite directions. There are two possibilities:

- the blood flows from the external carotid artery (reversed) via
  the internal carotid artery to the brain
- the blood flows from the internal carotid artery (stealing from
  the brain circulation - reversed) to the external carotid artery.

If by compressing the contra-lateral internal carotid artery the
flow increases in both vessels, then the former situation exists.
If by compressing the contra-lateral internal carotid artery the
flow decreases in both vessels, then the latter situation exists.
If no change occurs it is likely that the circle of Willis does not
function. In that case the vertebral artery HTG must be higher than
normal (mostly on the same side), especially if there is no hemi-
paresis. If compression of each of the vessels induces slow activity
in the electro-encephalogram, it is likely that the brain is supplied
with blood from the external carotid artery via the internal carotid
artery.

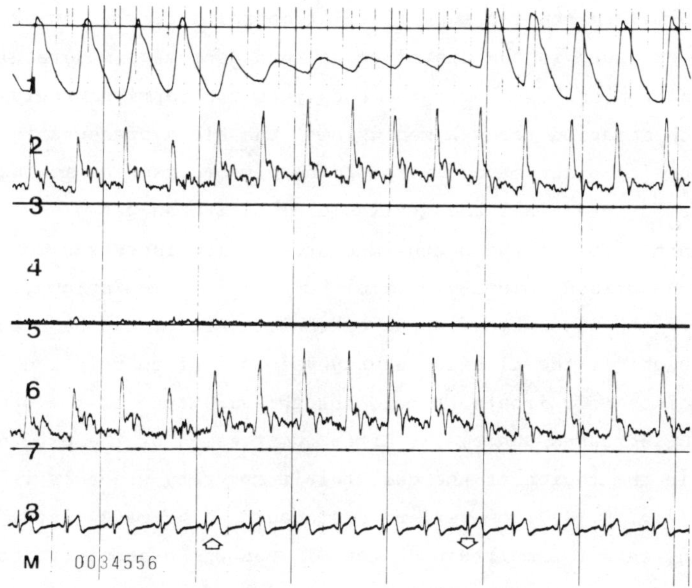

FIG. 6.10 HTG of the common carotid artery. Moderate
compression of the common carotid artery on the contra-lateral side (between arrows) results in a decrease of the amplitude of the plethysmogram on that
side, but in an increase in the amplitude of the HTG,
indicating that the left-to-right connection in the
circle of Willis is patent.

6.5.10 <u>Differentiation Between Total Obstruction Of The Common Carotid Artery Low In The Neck And Total Obstruction Of The Internal And External Carotid Arteries</u>

In both cases there is no or nearly no amplitude in the common carotid artery HTG. Sometimes the common carotid artery cannot be found at all. In case of a total obstruction of the internal and external carotid artery, the pulsations of the common carotid artery walls are strong, as can be palpated and demonstrated by placing the Doppler transducer perpendicular on the artery. If the obstruction is situated low in the neck there are, as a rule, no or nearly no pulsations in the common carotid artery. The bifurcation can be open or closed (see Section 6.5.9).

6.5.11 <u>Identification Of A Frontal Meningioma</u>

If in a normal person the superficial temporal artery is compressed, the blood pressure in the distal vessel bed decreases moderately. Several anastomoses take over the supply to the deprived region. If, however, this region feeds a meningioma, the blood pressure in the distal vessel bed decreases extra-ordinarily because the meningioma sucks - so to speak- all the available blood it can get.

In a normal person the ophthalmic artery flow increases moderately after the described compression. In the case of a meningioma, however, the rise of the blood flow in the ophthalmic artery can be 3 to 8 times as high. If the clinical neurophysiological correlation in a patient points to a frontal tumor, and the amplitude of the ophthalmic artery HTG increases abnormally after compression of the external branches in the region of the ear, this tumor very probably is a meningioma (Fig. 6.11). However, no statement can be made if the pronounced increase in amplitude of the HTG cannot be demonstrated.

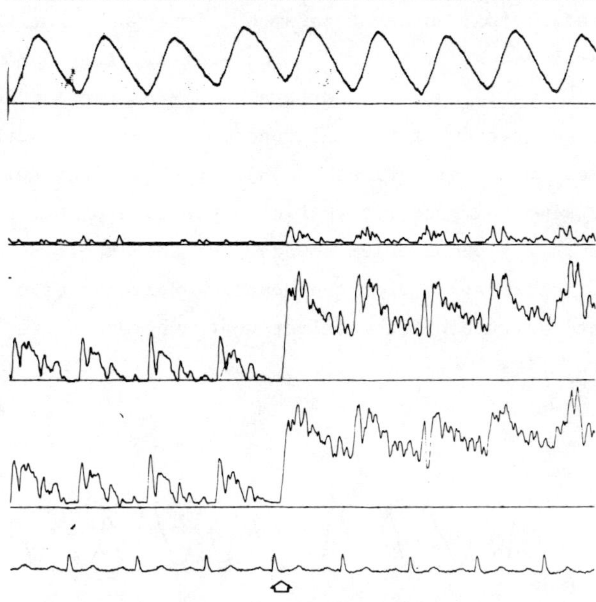

FIG. 6.11 HTG of an ophthalmic artery with normal amplitude and normal flow direction.
   After compression (arrow) of the supra auri-
   cular branches of the external carotid artery,
   supplying a frontal meningioma, the amplitude
   of the HTG increases considerably. The upper
   trace is the plethysmogram of the ipsi-lateral ear.

6.6   CEREBRAL DEATH

One of the usual findings in clinical death is the particular form of
the common carotid artery HTG. Following the S-wave a considerable
reflux is recorded. If one sees this phenomenon developing in the
course of time after a head injury the prognosis is very bad
(Fig. 6.12).

In cerebral death the flow in the internal carotid artery distal
to the bifurcation has to be zero. This means one does not succeed in
registering blood flow velocity in the internal carotid artery at all.
It is not easy to prove that there is no flow in the internal carotid
artery, because usually in this case the external carotid artery

shows a remarkable diastolic flow! In these patients, however, Doppler investigation can be of paramount importance for the diagnosis of cerebral death, if a skull defect exists, for instance after a trepanation. In all cases of cerebral death no Doppler sound can be heard if the transducer is positioned over the skull defect and slowly rotated in all directions. In all other patients with a skull defect it is almost impossible to find a position of the probe that does not result in loud Doppler sounds. For this application it is advisable (or necessary?) that the Doppler apparatus is provided with a switch that can turn off the filter that suppresses the very low Doppler frequencies.

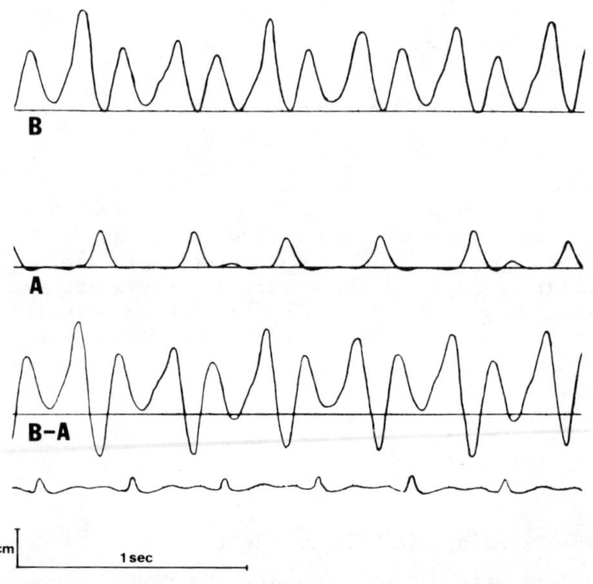

FIG. 6.12 HTG of the common carotid artery in a patient with a post-traumatic intracerebral hematoma, with the clinical signs of imminent cerebral death. Following the S-wave a considerable reflux is recorded as can be seen in the A and B-A channels.

6.7   SUBROUTINES IN THE DECISION PROCESS

As soon as the recording of the various HTG's has been finished a
Doppler diagnosis must be made. The patient should be kept in for
possible further investigations. Carrying out this, we examine the
HTG curves in a special sequence. In our department the following
subroutine in the decision process is used in evaluating the HTG
findings.

6.7.1   The Brachial Artery HTG

(a) Normal. Proceed to 6.7.2

(b) Abnormal - abnormally low S-wave bilaterally with or without diastolic retrograde or anterograde flow. There may be
some disease of the heart or generalised disease of
the arteries in which case the same abnormalities will
be seen in the femoral and other arteries. If the abnormalities are confined to both brachial arteries,
this may be due to Takayusu (pulseless) disease.

- increased anterograde flow in diastole bilaterally may
be due to decreased peripheral resistance in the arms
as a result of muscular exercise and other factors.
This can also be the case in the face. Then the external carotid artery might have a diastolic flow component also. The diastolic amplitude of the common carotid artery HTG does no longer reflect the internal
carotid artery flow alone.

- increased retrograde flow in diastole in both brachial
and other arteries may be due to aortic insufficiency
in which case recordings from the common carotid artery will show decreased anterograde flow during diastole. This decreased flow does not necessarily mean
stenosis.

- abnormally low S-wave unilaterally with prolonged rise
time and significant anterograde flow in diastole may
result from stenosis of the innominate or subclavian

artery. Check for subclavian steal syndrome and, if
in the right arm, check for stenosis of the innominate artery as well. Do the subclavian steal test
(see Section 6.4). Do the innominate artery test
(see Section 6.4).

- asymmetry with normal wave shape. Is the artery the
  brachial? (see Section 6.5.6). Do the brachial identification test. If it is the brachial artery look
  for stenosis in the vessels of the forearm.

6.7.2 The Ophthalmic Artery HTG

(a) Normal. Proceed to 6.7.3

(b) Abnormal   - abnormally low curve or reversed flow on one side
  probably indicates stenosis or occlusion of the ipsilateral internal carotid artery. To prove that the
  flow is indeed reversed do compression procedures and
  verify that the signal is not originating from the
  retro-ocular veins.

- abnormally high curve on both sides. There may be
  increased cerebrovascular resistance. Look for hypertension, biochemical disturbances, intracranial high-
  pressure.

  If the ophthalmic artery HTG is too high, then a
  low diastolic amplitude of the carotid artery HTG
  does not point to a stenosis localized between the
  bifurcation and the origin of the ophthalmic artery,
  but to a stenosis of the internal carotid artery
  beyond the origin of the ophthalmic artery. Exclude
  locally increased flow velocities in the bifurcation
  region and in the internal carotid artery.

- if the ophthalmic artery HTG is too high on one or
  both sides with normal flow direction and the amplitude of the carotid artery HTG during diastole is
  normal on that side, but with a low amplitude during
  systole, one should consider a stenosis of the external carotid artery or its branches on that side.

Look for local increases in flow velocity at the
bifurcation and in the external carotid artery.

6.7.3   The Common Carotid Artery HTG

(a) Normal. Proceed to 6.7.4

(b) Abnormal - abnormally low S-wave bilaterally with good S/D index
may be due to cardiac disease in which case the brachial arteries will show the same features. However, if
the brachial artery HTG is normal, a bilateral stenosis
may exist low in the common carotid artery, especially
if the steepness of the S-wave is less than normal.

- abnormally low D-amplitude with increased S/D index
bilaterally with normal brachial artery HTG may indicate bilateral stenosis of the internal carotid arteries. Try to find high flow velocities and turbulence
at the bifurcation as a reflection of local stenosis.

- a diastolic asymmetry greater than 25% indicates disease on the side of lower amplitude. If the sum of the
D values of the two sides is approximately twice normal,the stenosis or occlusion of the internal carotid
artery on the abnormal side is being compensated by
the Circle of Willis. If the sum of the two D values
is less than the sum of two normal values there may be
a lesser degree of stenosis also present in the "normal" internal carotid artery.

- a flattened S-wave may result from turbulence in which
case the S/D index and the resistance index of Pourcelot will not be reliable.

- the absence of anterograde diastolic flow on one side
may be due to total occlusion of the ipsi-lateral internal carotid artery. When considerable retrograde
ophthalmic artery flow is found, a (low) anterograde
diastolic flow in the common carotid artery can be
expected. Try to find high frequency Doppler sounds
in the bifurcation region. If these are present the
obstruction is not total yet.

- retrograde diastolic flow on both sides with a normal
  or high brachial HTG indicates a high cerebral vascular resistance such as occurs with markedly increased
  intracranial pressure and cerebral death.
- the absence of the HTG on one side may be due to proximal occlusion of the common carotid artery or occlusion of both internal and external carotid arteries
  near the bifurcation. Do the pulsation test of the
  common carotid artery (see Section 6.5.10). If there
  are hardly any pulsations, try to find out if the
  bifurcation is open. Do if indicated, the contralateral compression test to determine, whether there
  is a steal from the brain.
- the absence of the HTG bilaterally may be due to bilateral common carotid or bilateral external and internal carotid artery occlusions in which case the whole
  of the cerebral circulation will be arising from the
  vertebral arteries with corresponding increase in
  their HTG amplitudes.

6.7.4 The Vertebral Artery HTG

(a) Normal. The examination is concluded

(b) Abnormal - the S-wave is abnormally high on one or both sides
    when the vertebral artery flow is compensating for
    reduced carotid flow or for increased flow to an ar-
    terio-venous malformation in the territory of the
    vertebral-basilar system.
- the direction of flow is totally or partially reversed
  in subclavian steal syndromes (see Section 6.4)

6.8 CLINICAL NEUROPHYSIOLOGICAL CORRELATIONS

After having made the "Doppler diagnosis" one must see how the Doppler
findings fit to a known clinical neurophysiological syndrome. These
syndromes are well-known combinations of case history, clinical

findings as well as the findings on the EEG, Echo-encephalography
and Doppler HTG.

A few examples are:
1. The syndrome of an internal carotid artery stenosis on one side
   which is, most of the time, compensated.

CASE HISTORY:	once or a few times blurred vision and transient ischemic attacks on the other side.
CLINICAL FINDINGS:	no abnormalities
EEG:	normal
ECHO:	normal
HTG:	stenosis of the internal carotid artery on the contralateral side of the paresis
DECISION:	angiography + operation

2. The syndrome of the uncompensated internal carotid artery stenosis
   on one side.

CASE HISTORY:	unilateral paresis
CLINICAL FINDINGS:	unilateral paresis
EEG:	contralateral severe EEG abnormalities
ECHO:	normal
HTG:	stenosis of the internal carotid artery contralateral to the paresis
DECISION:	angiography and operation after clinical recovery

3. The syndrome of a relatively small ischemic lesion of the capsula
   interna.

CASE HISTORY:	unilateral paresis
CLINICAL FINDINGS:	unilateral paresis (hypertension?)
EEG:	normal or nearly normal
ECHO:	normal
HTG:	normal
DECISION:	no angiography. CT-scan. Conservative treatment.

   If there has been high blood pressure and an abnormal HTG on the other side, it probably would have been a capsula bleeding on one side and a carotid artery stenosis on the other side. Angiography should be considered, with all the risks involved.

When writting this chapter, we have had the experience of nearly 10 of such correlation syndromes. It is the task of the clinical neuro-physiologist and the neurologist to determine which syndrome is applicable to the patient in question. On the basis of these syndromes decisions have to be made for further investigation and treatment. It is clear that a wrong HTG registration, or a wrong interpretation may have far reaching consequences.

LITERATURE RELATED TO THE SUBJECT

Fields, W.S. (1965). Collateral circulation of the brain. The Williams and Wilkins Comp. Baltimore.

Mol, J.M.F. and Rycken, W.J. (1974). Doppler haematotachographic investigation in cerebral circulation disturbances. In Cardiovascular applications of ultrasound (Edited by R.S. Reneman).pp. 305-315. North-Holland/American Elsevier, Amsterdam-London-New York.

Pourcelot, L. (1976). Diagnostic ultrasound for cerebral vascular diseases. In Present and Future of Diagnostic Ultrasound (Edited by I. Donald and S. Levi). pp. 141-147, Kooyker Sci., Rotterdam.

Reneman, R.S. and Spencer, M.P. (1979). Local Doppler audio spectra in normal and stenosed carotid arteries in man. Ultrasound in Medicine & Biology, 5, 1-11.

Reutern, G.M. von and Pourcelot, L. (1979). Cardiac cycle-dependent alternating flow in vertebral arteries with subclavian artery stenoses. Stroke, 9, 229-236.

Spencer, M.P., Brockenbrough, E.C., Davis, D.L.,and Reid, J.M.(1977). Cerebrovascular evaluation using Doppler C-W ultrasound. In Ultrasound in medicine 3B (Edited by D. White and R. Brown). pp. 1291-1310. Plenum Press, New York-London.

Zülch, K.J. (1971). Some basic patterns of collateral circulation of cerebral arteries. In Cerebral Circulation and Stroke (Edited by K.J. Zülch). pp. 106-122, Springer Verlag, Berlin-Heidelberg-New York.

CHAPTER 7
Using Carotid Imaging and Hand-Held Probing Doppler Evaluation of the Aortocranial Circulation

M. P. Spencer

7.1 INTRODUCTION

The purpose of this chapter is to present an integrated approach to diagnosis of lesions in the extracranial arterial supply to the brain relying on newly available full capability instrumentation which presents all of the features of Doppler ultrasound including imaging of the carotid artery bifurcation. A complete disclosure of the basic principles of continuous wave (CW) Doppler examination has been published elsewhere (Spencer and Reid, 1981). Material discussed here is supplemental and discloses newer techniques of diagnosis and results.

The overall objective of Doppler cerebrovascular diagnosis is to prevent stroke and symptoms of cerebrovascular insufficiency by providing cost-effective non-invasive diagnoses which provides information for clinical management decisions. For successful application of Doppler to the problem of stroke prevention, a wide range of expertise is necessary including an understanding of:

- instrumentation operation and limitations
- technique of patient examination
- pulsatile hemodynamics
- interpretation of signals in various audio-visual presentations
- the integration of Doppler findings with the patient's history, physical findings and other laboratory tests
- treatment options available and their suitability for the patient.

The complete cerebrovascular evaluation in our laboratory is outlined in Table 7.1. See sample report at end of this chapter.

TABLE 7.1 : Major elements of our present non-invasive cerebral
vascular evaluation

History:

    Carotid insufficiency symptoms
    Vertebrobasilar symptoms

Physical examination:

    Neurological survey
    Arterial palpation
    Auscultation of hand, neck and upper chest
    Bilateral brachial artery pressures

Doppler examinations:

    Vertebral artery signals
        Base of skull (Atlas loop)
        Anterior-supraclavicular
        Subclavian, auxillary and brachial artery signals
    Hand-held probing
        Ophthalmic artery signals
            Posterior orbital
            Periorbital
        External and internal carotid signals
    Doppler imaging
        Carotid bifurcation
        Subclavian-vertebral and low common carotid arteries

Diagnostic interpretations:

    Carotid arteries
        Bifurcation - stenosis or occlusion (of internal,
        external or common carotid arteries)
        Aneurysm
        Collateral evaluation
        Plaquing without stenosis
        Siphon stenosis
        High intracranial resistance
    Vertebral arteries
        Stenosis
        Subclavian steal
        Carotid Collateral Function
    Subclavian and innominate arteries
        Stenosis and occlusion
    Cardiac
        Low output
        Arrhythmias
        Miscellaneous

In the clinical report the findings of the physical examination and the Doppler examination should be separated from an interpretation of their patho-physiological meaning of the findings. Finally a consultative recommendation should be given considering the patient's symptoms, age and other conditions, making recommendations concerning X-ray angiography and surgery, antiplatelet medication and anti-coagulants, following the patient with no special treatment, or other test procedures such as echocardiography or EEG.

7.1.1 Continuous Wave (CW) Doppler Ultrasound

The general features of continuous wave Doppler ultrasonic veloci-meter for transcutaneous use has been described in other chapters of this book. The special features of the CW Doppler device as origi-nally developed in our laboratories, incorporate a net directional flow feature and two selectable frequencies to operate a lens focus-sing 5 MHz probe, converging at a depth of 2.3 cm with a beam dia-meter of 1 mm and a smaller 10 MHz "pencil" probe for more superfi-cial and smaller arteries. By means of simple replacable crystals the equipment accommodates probes varying in frequency between 1 and 10 MHz (Fig. 7.1). Recent changes in electronics provide for dual-directionality using the circuit of Nippa et al (1975), recordings for the interpreting physician included audio tape recordings and strip chart tracings $A^2\omega$ (A = amplitude; ω = frequency, Reid et al, 1974).

Continuous wave Doppler ultrasound was chosen because of its ad-vantages over pulsed Doppler in the assessment of the peripheral cir-culation. These advantages include:

- a better signal-to-noise ratio
- higher frequency response for detection of high blood velocities
- easier vessel identification because of a broader ultrasonic beam and no necessity to range the depth of the vessel
- a greater cost/effectiveness ratio.

Reneman and Hoeks (1977) have disclosed the pulsed Doppler trade offs between frequency response and pulse repetition rate.

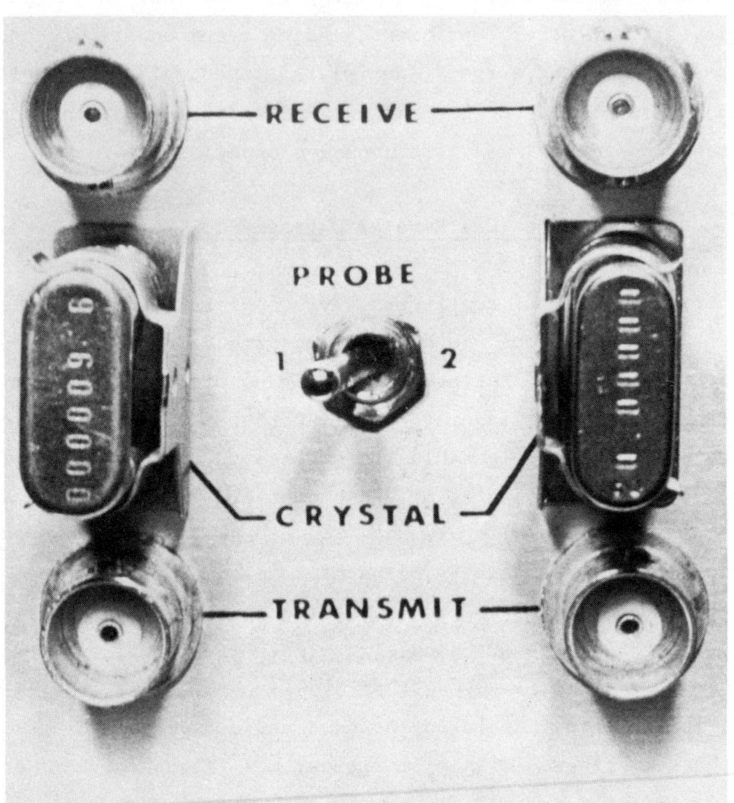

FIG. 7.1   The front panel of the Doppler module illustrating dual capacity for two probes. The switch is in
position 1 to operate the receive and transmit crystals
on the left. The interchangeable crystals to maximally
tune the probe crystals is seen in the holder clips
between receive and transmit connectors.

The potential advantages of pulsed Doppler include more positive
identification of deeper arteries such as the vertebral and intra-
thoracic segments of the aortic root branches, higher resolution of
morphology such as vessel diameter, luminal craters, detailed study
of velocity profiles, and as an assistance to real-time pulsed echo
imaging. A hope for the future which pulsed Doppler may achieve is

the interrogation of individual intracranial arteries but the higher
power of pulsed Doppler required for detection of deep vessels
through the posterior orbit may provide a risk to the eye.

The principle contribution of CW Doppler to cerebrovascular evaluation is to identify internal carotid artery stenosis non-invasively.
Table 7.2 discusses the results of comparison between X-ray angiography and our original CW Doppler to diagnose carotid artery obstructions. This comparison was performed using analog tracings and listening to the audio signals without spectral analysis. Additional
capability of Doppler to diagnose subclavian and vertebral artery
abnormalities is less well recognized.

TABLE 7.2 : Non-invasive cerebrovascular evaluation vs. X-ray angiography. Experience of Davis and Gingery (personal communication):
16 months, 133 patients, 257 carotid arteries

Test	% stenosis Angio	Sensitivity	Specificity	Diagnostic Accuracy
Ophthalmic direction	$\geq$ 70	26	80	66
Phonoangio	$\geq$ 50	59	82	74
Periorbitals	$\geq$ 70	49	90	83
DOPSCAN imaging	$\geq$ 50	90	84	86
Combination	$\geq$ 60	94	94	94

7.1.2 Features Of Continuous Wave Doppler Signals

All Doppler signals possess five usable features, all of which should
be used to the utmost in diagnosing extracranial arterial disease:
- amplitude
- frequency distribution
- phase or direction of flow
- pulsatile variations
- location or source.

All Doppler signal characteristics can be modified by interventions
such as mechanical compressions or pharmacological means to assist
the examination and its interpretation. The total amplitude is the
least reliable feature because it depends on many factors which do

not relate to blood velocity or blood flow. The distribution of
amplitudes or power distribution within the frequencies of the spectrum is, however, a valuable diagnostic feature.

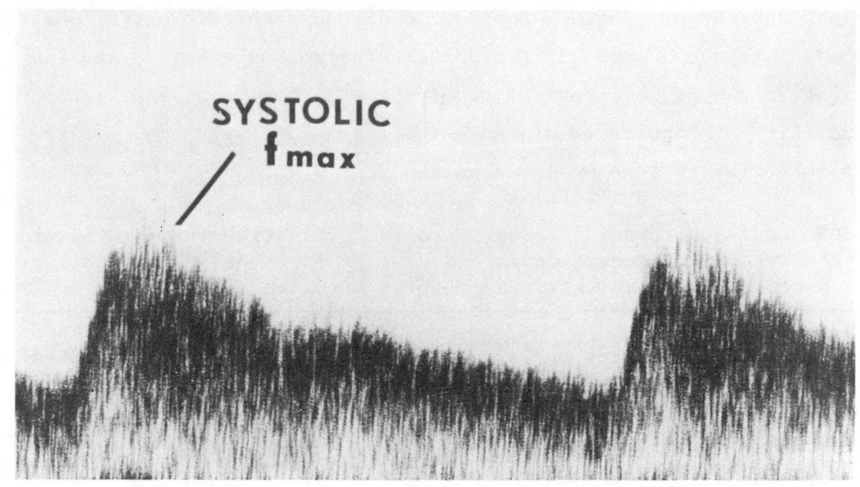

FIG. 7.2   Typical spectrum from normal internal carotid artery
of a human subject demonstrating the concentration of energy
at the upper edge of the pulse wave. f_{max} is defined as the
frequency at the upper edge of the spectrum. Peak systolic
f_{max} is indicated.

Maximum frequency or the upper edge of the spectrum (Fig. 7.2) is
the most usable feature particularly when following the signal along
the axis of vessels, when comparing symmetrical arteries on opposite
sides of the body, and in following changes within the same artery
from one examination to the next. Because blood velocity varies

across the vessel, the display of spectral distribution as function
of time is of great diagnostic value. Many new audio spectral analyzers have become commercially available in 1980. Their price
(approximately $10,000) now makes their routine clinical use cost
effective for an active vascular laboratory. The mean spatial frequency is often computed using the zero-crossing meter. Its advantage
is a simplified read-out of mean velocity waveform. The velocity frequency distribution of the signal is reduced to an average frequency
(Fig. 7.3). When gross turbulence is present its accuracy is compromised.

The direction feature of the blood flow is usually identified by
the phase sense of the Doppler shift. Reliability of this feature is
occasionally compromised by proximity of bony structures or a curvature of the artery direction. An effort should be made to know and
control the angle of the ultrasonic beam with the flow axis. Direction is defined in our laboratory in various ways: "forward" moving
in the normal direction towards the tissues; "reversed" being abnormal arteries (aphysiological as defined by Mol in chapter 6); "inverted" meaning polarity is exchanged, appearing to be reversed when
in fact calcium deposits have artifactually modified the apparent
polarity; "bidirectional" meaning at the same approximate location
the signals are obtained with either positive or negative polarity.
"Biphasic" means the signal direction changes within the cardiac
cycle (Fig. 7.4) and "dual direction" refers to flow streams within
the ultrasound beam moving simultaneously in both directions, e.g.
in turbulence.

Useful audio characteristics have been verbally defined; "high"
meaning the maximum frequency is higher than normally detected in
vessels at that location; "low" meaning lower frequencies than usually obtained from that arterial location; "smooth" meaning the spectral distribution of frequencies is concentrated near the maximum
frequency throughout the heart cycle (Fig. 7.5A), "coarse" meaning a
spectral broadening, which is usually not detected at that location
in normal arteries (Fig. 7.5B); "gruff" meaning low frequency accen-
tuation in systole with frequencies between 100 and 500 Hz (Fig. 7.6B)
(these are usually of large amplitude); and "fluttering" is a special

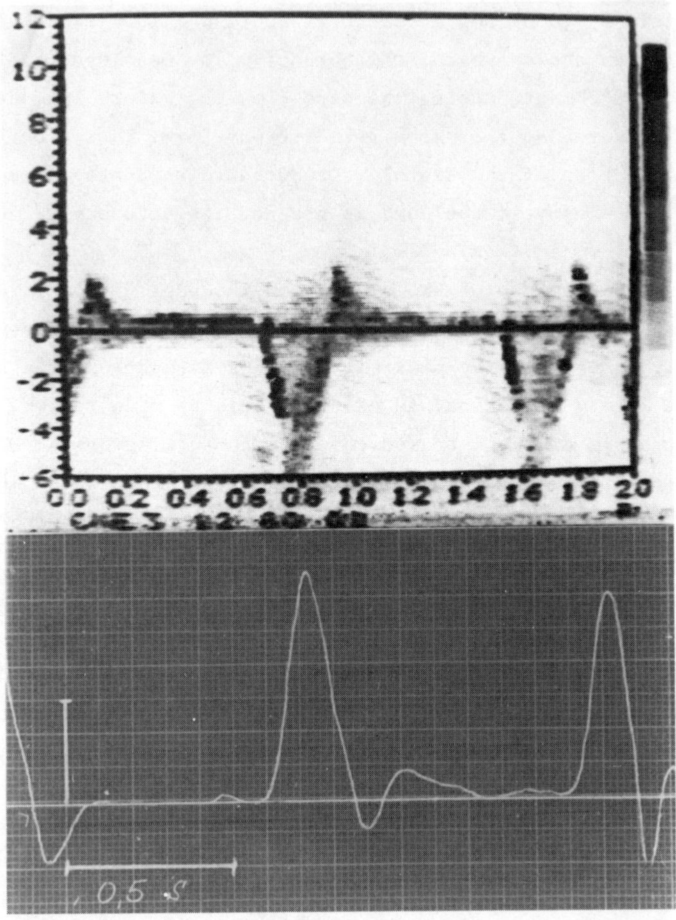

FIG. 7.3   Comparison of spectral analysis waveforms in the upper panel with the zero-crossing
   meter output analog tracing of the mean frequency in the lower panel. The signal for both
   tracings is taken from the normal brachial
   artery of a human subject. (Courtsey of Dr.
   G.M. von Reutern, Freiburg, Germany)

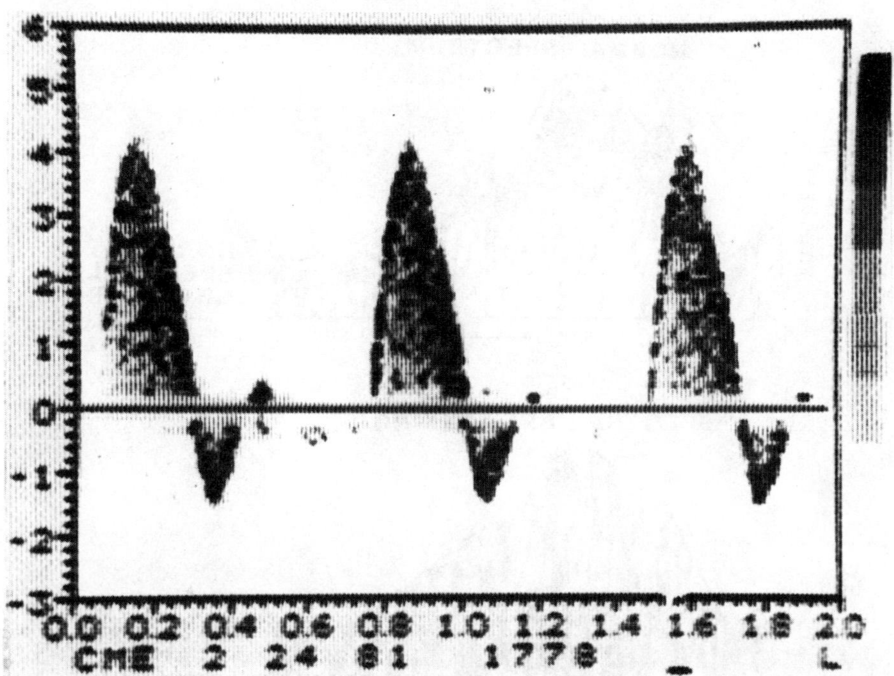

FIG. 7.4   Black and white print of color transparency taken of a video presentation of the dual-directional color spectral ana-lyzer (SONACOLOR). Note changes in the spectrum reflecting the differences in the acceleration profile and the deceleration profile. Concentration near the maximum edge is more evident during acceleration because the velocity profile is more flat during acceleration than during deceleration. Coordinates in kHz and seconds.

audio impression seen on the maximum edge of the spectrum as a spiking irregularity (Fig. 7.6A). This, in a more severe state termed "stuttering" (Fig. 7.7A) represents turbulent flow. A whining or "sea gull" quality heard at 100 and 500 Hz (Fig. 7.7B) caused by a periodic oscillation of the artery wall where a blood jet stream emerging from a stenotic segment grazes the vessel wall. Gruff signals are detected downstream to stenosis (Reneman and Spencer, 1979) and represent the bruit caused by random vibration of the artery wall due to underlying high energy turbulence.

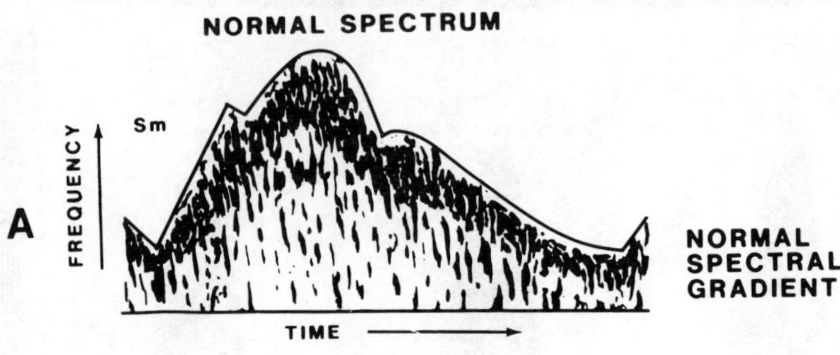

FIG. 7.5   Diagrammatic representation of spectral appearances of
the internal carotid artery for the normal spectrum in panel A
and with spectral broadening in panel B. Spectral broadening provides a coarse (C_o) quality to the sound as opposed to smoother
(S_m), more "whistling" quality of a normal spectrum.

Pulsations of Doppler variations within the cardiac cycle carry
a great deal of physiological information for the interpreter who
understands the underlying hemodynamics. Definitions include "systolic" (high pulsatility) which means blood flow primarily during the
systolic phase of the arterial pulse, "continuous" (low pulsatility)
meaning a high diastolic flow component; "backflow" meaning a reversed flow at the end of systole or in early diastole; "rapid or slow
acceleration" referring to the rate of change of the velocity or
frequency particularly noted on the leading edge of the pulse;
"damped" signal also is a term associated with slow acceleration.
"Phase lag" is a term applied to the delay in the pulse as compared
with a reference signal such as the ECG or another Doppler signal.

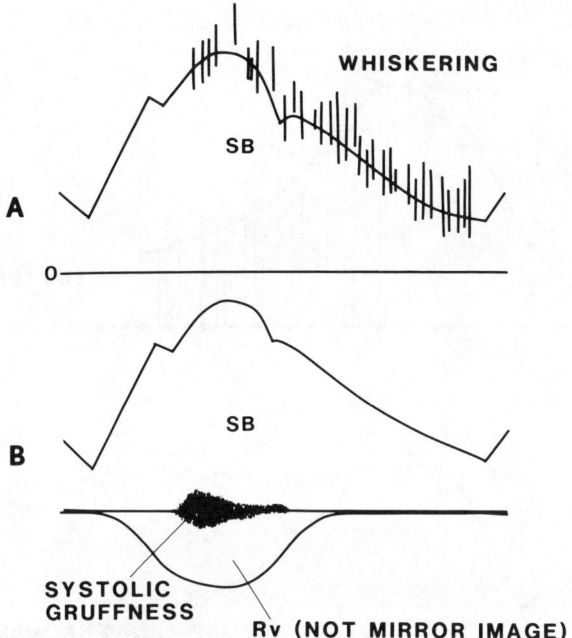

FIG. 7.6   Diagrammatic representation of
two spectral abberations produced by
turbulence. Panel A illustrates "whiskering" irregularity to the upper edge
of the spectrum. This is usually accompanied by spectral broadening (SB).
Whiskering generally begins in mid-systole at peak velocities when turbulence
is minimal. Often, this sign of turbulence appears before a bruit can be
heard over the arterial channel. Panel
B diagrams the appearance of the "gruff"
systolic qualities which represent the
bruit on the Doppler signal. There is
a concentration of high energy around
the zero base line, usually less than
1,000 kHz, more apparent during systole than during diastole. In addition
the reversed (Rv) spectrum during systole with spectral broadening is typical of rotating vortices in flow streams
downstream to a segmental stenosis.

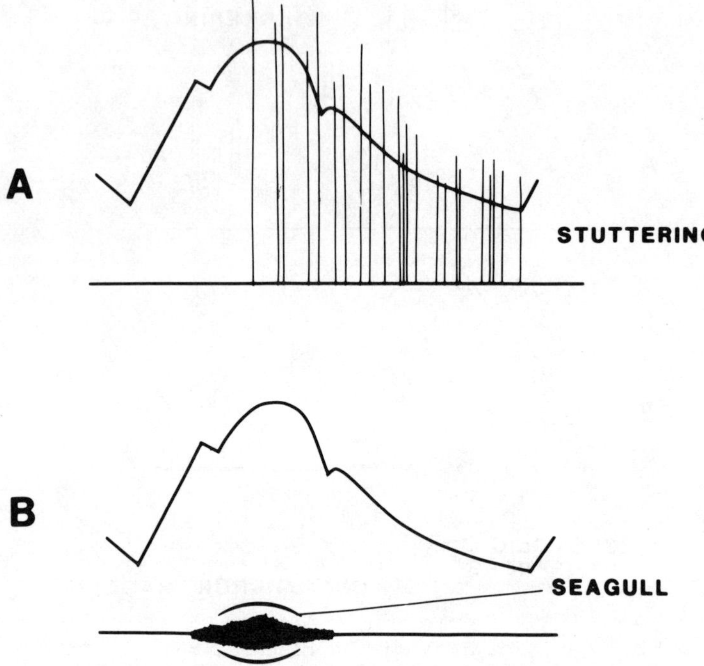

FIG. 7.7   Panel A - Stuttering which represents a
violent form of turbulence in which whiskering on
the maximum edge extends to the base line of the
spectrum. This stuttering probably represents an
oscillation in direction of the downstream jet,
moving back and forth across the artery lumen.
Panel B represents a "sea gull" or whining, sometimes moaning, quality accompaning the Doppler
representation of the bruit. This phenomenon
probably is caused by grazing of the jet close
to one side of the downstream artery wall, causing it to oscillate in a periodic fashion.

The location or source of the Doppler signal is all important in identifying the underlying vessel and specifying the location of disease. Doppler imaging techniques represent a special way of more precisely identifying the location of abnormal signals (Spencer et al, 1974a). CW Doppler imaging (DOPSCAN[*]) for the carotid and femoral artery bifurcations is discussed later in detail. The technique of hand-eye imaging of the carotid arteries should be noted. This method of "cortical" imaging has been developed to a high skill by Von Reutern et al (1976b) and is used in Europe by many skilled physicians (Fig. 7.8). It is essentially a hand-held method of localization of signal sources coupled with listening analysis of the signals. Auxiliary analog tracings are used.

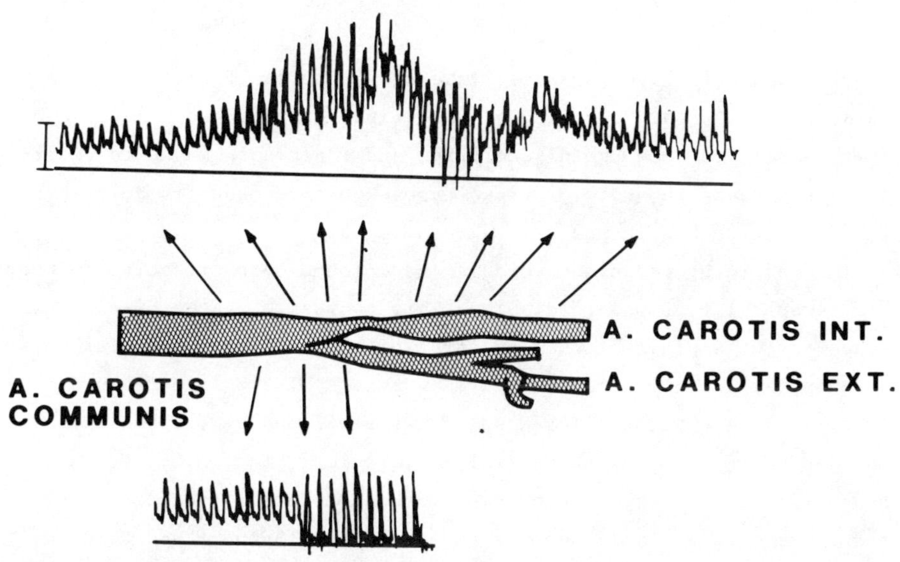

FIG. 7.8   Analog zero-crossing tracings on "run-throughs" of common-internal, above and common-external, below. The upper tracing illustrates progressively increasing frequency as the sound beam enters the stenotic segment on the origin of the internal carotid artery. Downstream to the stenosis, analog tracings become confused because of the predominance of the low frequency information. At the furthest extremity, the analog tracings return to a better representation of the mean frequency. Diagnostic features of stenosis, however, are represented in graphic form (techniques of Von Reutern).

[*]Manufactured by Carolina Medical Electronics, King, NC, USA

7.1.3 Audio-visual Displays

Many audio-visual presentations of Doppler signals assist the interpreter, but listening with the trained human hearing mechanism is of great value because all signal features are represented including direction, if stereo speakers are used, and the human hearing mechanism is capable of complex audio recognition of most diagnostic characteristics. It lacks somewhat in objectivity and quantitation, but notations using the definitions just given, can be made of the signal qualities to provide documentation. Listening remains the single most useful analysis technique.

Documentation of the signals by spectrum analysis graphing, analog read-outs or prints of imaging assists the interpreter by teaching how and where to recognize the important audio variations and provides quantitative parameters and documentation for clinical records and research. Reduction of the signals to hard copy may, however, eliminate some useful qualities and replayable magnetic recordings such as with audio or video tape should be made for delayed interpretation and re-analysis when necessary.

The recording technique which we have found most effective we term "full capability Doppler" display. This composite system displays in video format a bidirectional spectrum of bidirectional audio signals and Doppler imaging or other representations of the signal source (Fig. 7.9). The video color display was developed by Dr. Ron Hileman and Joe Cairo of Carolina Medical Electronics, King, N.C., U.S.A., and utilized a microprocessor single-board computer and fast Fourier transform with a 5 milliseconds updating time (Spencer and Reid, 1980). It operates on the phase difference between signals generated from motion toward the ultrasound source and motion away from the source. Blood velocities away from the probe are represented by frequencies above the zero base line while blood velocities towards the probe are represented by frequencies below the base line. Three selectable frequency scales are available on the ordinate including 0 to 6 kHz positive and 0 to 3 kHz negative; 0 to 12 kHz positive and 0 to 6 Hz negative; 0 to 20 kHz positive to 0 to 10 kHz negative. Three time bases are selectable including full sweeps of 1, 2 or 10

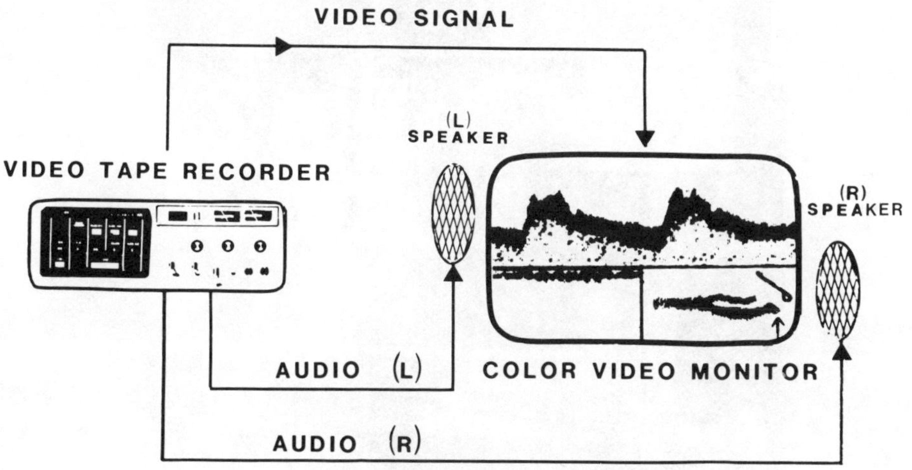

FIG. 7.9   New full capability technique for audio/visual display of
   diagnostic Doppler signals. The video monitor displays the spectrum of the signal arising from a position indicated by a cursor
   on the image quadrant of the screen. Directional speakers are
   provided in addition.

seconds. The amplitude of the audio frequencies is represented by co-

lor coding using a thermal scale with red representing the lowest

amplitude and white the greatest amplitude. Oranges and yellows re-

present the middle ranges. A 16 level grey scale is also available.

An automatic thresholding technique increases the speed of imaging and diagnostic accuracy. We are currently evaluating its total impact on non-invasive diagnosis of stroke conditions.

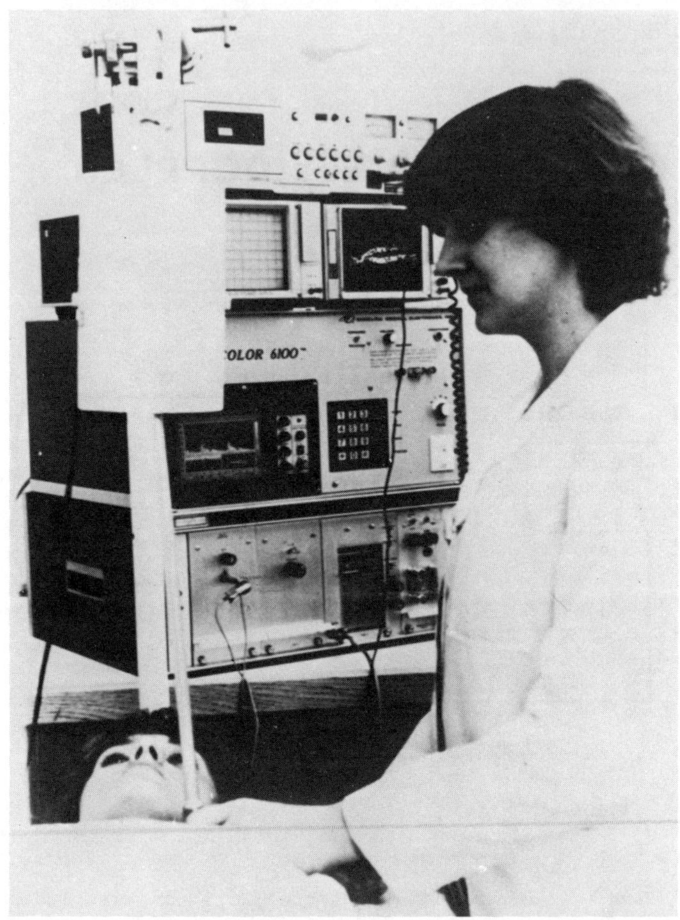

FIG. 7.10   Arrangement of instrumentation, patient and
technician for use of Full Capability Doppler display.
Special screen at the technician's eye level displays
the carotid bifurcation image in life-size. SONACOLOR
spectral analyzer seen at neck level of the technician
displays the spectrum as well as Full Capability Doppler
representation on video. The probe positioning arm extends vertically over the supine patient. The fulcrum of
the arm, swinging in the lateral direction, is over the
patient's mid-line. This allows approximately a 5 degree
angle to partially separate the external from internal
carotid artery image projection.

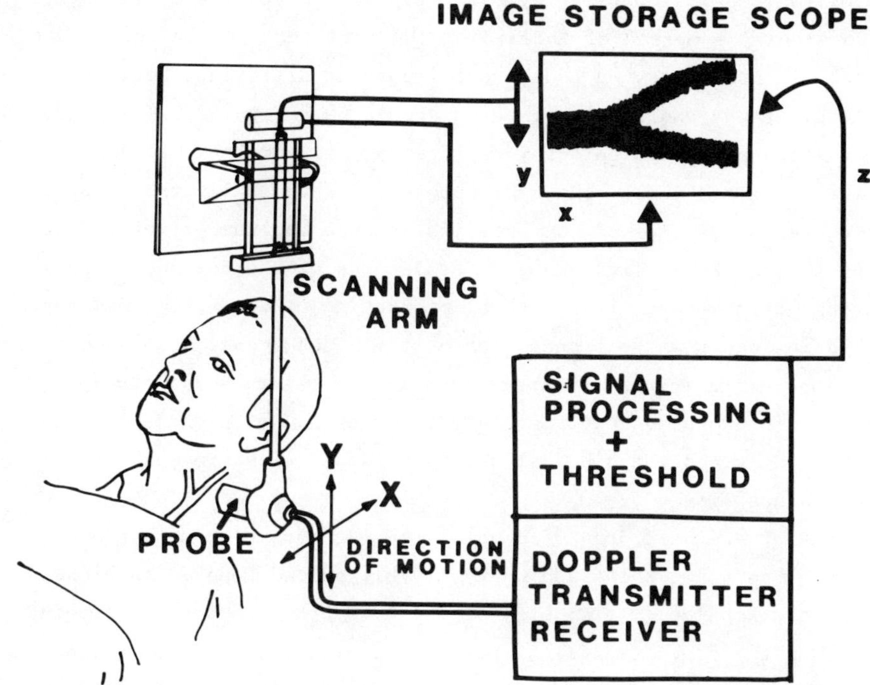

FIG. 7.11   Diagrammatic representation of the continuous-wave
Doppler imaging technique.

7.1.4   Doppler Imaging

The CW Doppler imaging instrumentation (DOPSCAN) is illustrated in
Fig. 7.10 where the equipment is arranged for imaging the carotid
bifurcation. Fig. 7.11 is a diagrammatic representation of the functional aspects of the imaging instrumentation. Originally the image
of the carotid bifurcation was recorded on fast film. Presently, we
use video tape recordings of all audio and imaging information diagrammed in Fig. 7.9. During the examination the technician monitors
the audio-frequencies by means of dual directional speakers and observes the spectral display. Voice annotations are shared with one
of the Doppler audio channels. The most recent instrumentation also
displays ECG or ear-pulse signals.

174

Neither continuous wave nor pulsed Doppler imaging displays the morphological resolution of X-ray angiography. The Doppler image should be considered a method of localizing the pertinent physiological and diagnostic qualities. An additional important difference between X-ray angiography and Doppler imaging should be recognized. The primary information presented on X-ray angiographic film relates to diameters of the projected lumen. The walls projected are usually the anterior and posterior aspects of the arteries because the numbers of projections available are limited by the tolerance of the patient. Doppler imaging, by contrast, is a lateral projection of the velocities represented in the cross-sectional area of the blood vessel with some modifications of the signal by aberations in the vessel wall. The interpretation of Doppler requires analysis of the audio signals. The Doppler imaging without audio analysis is not sufficient to make a reliable diagnosis.

Three other variations of Doppler imaging have been marketed in the USA. CW (Echoflow; White and Curry, 1978), pulsed Doppler (Angioscan), and pulsed Doppler (Mavis; Fish et al, 1978). Pulsed Doppler imaging relies on the depth (range) of resolution of the pulses to provide cross-sections of the artery flow channels. Cross-sections are made after producing the plane projection. Pulsed ultrasound cross-sections have better morphological resolution than CW ultrasound but require additional time. The primary signal of stenosis, high frequency of increased velocity, is compromised by the limited response of pulsed Doppler. Echoflow provides a color representation of three frequency categories. These categories use arbitrary frequency levels derived from the audio signal. The assigned frequency bands, for example, may be greater than 5 kHz, 3.5 to 5 kHz, and 600 Hz to 3.5 kHz with the highest band representing stenosis. While some success has been achieved by color coding in detecting moderate grades of stenosis the method by limiting its frequency resolution has problems like those of pulsed imaging in resolving severe stenosis greater than 90% and less than 100%.

The principle advantage of Doppler imaging of the arterial channels over hand-held probing relates to its ability to localize the source of the signals. This is of particular advantage at the carotid bifurcation where hand-held probing may not insure localization

within the important first centimeter of the internal carotid artery.
Stroke-producing plaques are prone to develop at this location but
lesions may be confused with those on the external carotid artery.
Careful searching for diagnostic signals at those crucial locations
is facilitated by the image. With Doppler imaging the angle of the
probe sound beam with the vessel axis is fixed and not subject to
hand-held variations. By this means greater reliance on changes of
frequency and other features greatly assists the localization and
diagnosis efforts. When inverted signals or non-sounding segments
produced by calcified plaques are present hand-held probing may be
confusing.

7.1.5   <u>Severe Stenosis vs Occlusion Of The Internal Carotid Artery</u>

With direct CW Doppler imaging the very important 90% pre-occlusive
stenosis (less than 0.5 mm in diameter and often without bruit), can
be differentiated from a total occlusion of the internal carotid artery. This difference with many experts determines whether carotid
endarterectomy will or will not be performed. This is a crucial clinical matter in stroke prevention, not because of the small difference
in brain perfusion which may occur upon occlusion, but because of the
danger of major stroke by embolization from red fibrin clots forming
downstream to the stenosis. In these "pre-occlusive" degrees of stenosis the velocity in the internal is slowed to such an extent that
large thrombi form and may be dislodged by the remaining flow through
the stenotic channel. It seems probably that many strokes that occur
at the time of total occlusion are caused by this mechanism and not
by the final loss of a small amount of flow in the internal carotid
artery. It is certain that the diagnostic technique relying on secondary features such as the pulsatility index and on tertiary collateral features of carotid artery obstruction cannot differentiate the
severe grades of stenosis from total occlusion. Hand-held Doppler
probing of the internal is likely to miss the weak signals found by
Doppler imaging. Table 7.3 discloses the ability of our original CW
Doppler techniques (without spectral analysis) to distinguish between
severe stenosis and total occlusion.

The objective of this investigation was to evaluate, by means of X-ray contrast angiography, the ability of non-invasive Doppler ultrasonic imaging of the carotid bifurcations to separate tight stenosis from total occlusion of the internal carotid artery. We used only audio analysis without the benefit of on-line spectral analysis which became available only during 1980. The population included all patients angiographically diagnosed at Providence Medical Center, Seattle, Washington during 1979 having less than or equal to 1.0 mm remaining lumenal diameter and who were evaluated at the Institute of Applied Physiology and Medicine non-invasive vascular laboratory at Providence Medical Center. Doppler interpretation was based on the techniques listed in table 7.1 listening to all Doppler signals without benefit of spectral analysis.

Ninety patients underwent cerebrovascular evaluation by both Doppler ultrasound and X-ray contrast angiography. Twenty-six of these patients displayed a total of 31 internal carotid artery lesions that were angiographically measured as being less than or equal to 1.0 mm. Five patients demonstrated bilateral lesions. Nine were totally occluded and 22 patent but tightly stenosed.

Doppler interpretation revealed 7 true positives, 21 true negatives, one false positive, and 2 false negatives. The over-all accuracy was 90.3% (28/31), while the sensitivity was 77.8% (7/9), and specificity was 95.5% (21/22).

The one false positive finding correlated with a patient that was reported by angiographers to have a stenosis of less than 1 mm angiographically located at the "bony canal through the temporal bone". The unusual location of this lesion above the region of bifurcation imaging accompanied Doppler evidence of external to internal carotid artery collateralization via the periorbital and posterior ophthalmic arteries. This condition probably resulted in velocities below that which is detectable by Doppler ultrasound and represented a misinterpretation of Doppler data. Analysis of the two false negative cases revealed that the non-invasive Doppler cerebrovascular evaluation was performed two months and 1.5 years, respectively, prior to angiography. Both of these cases are included because we set no time limit on the interval between the non-invasive study and angiography. However, it is probable that in both of these cases progression of disease accounted for the false findings because the primary features of stenosis (high local velocity within the stenosis) was present at the time of both Doppler evaluations.

Other accuracy studies made before, both with and without off-line spectral analysis have been previously published (Spencer et al, 1974a; Spencer and Reid, 1979).

TABLE 7.3 : The ability of Doppler ultrasonic imaging to separate tight stenosis from total occlusion of the internal carotid artery. A comparison with X-ray angiography

X-RAY ANGIOGRAPHY

	+	−
+	7	1
−	2	21

DOPPLER IMAGING

+ = occluded
− = patent

SENSITIVITY : 77.8% (7/9)
SPECIFICITY : 95.5% (21/22)
ACCURACY   : 90.3% (28/31)

NUMBER OF PATIENTS : 26
TOTAL ARTERIES   : 31

7.1.6   Carotid Artery Siphon Stenosis

This lesion must be diagnosed from primary high frequency Doppler features because of the inaccessability of this segment of the artery within the cavernous sinus. It can, however, be suspected from manifestations of secondary and tertiary effects. A bruit over the homolateral eye is a frequent tip-off raising suspicions. An increase in the internal carotid artery pulsatility index especially with reference to the opposite carotid artery or tertiary effects of collateralization in the absence of bifurcation stenosis also assists this diagnosis.

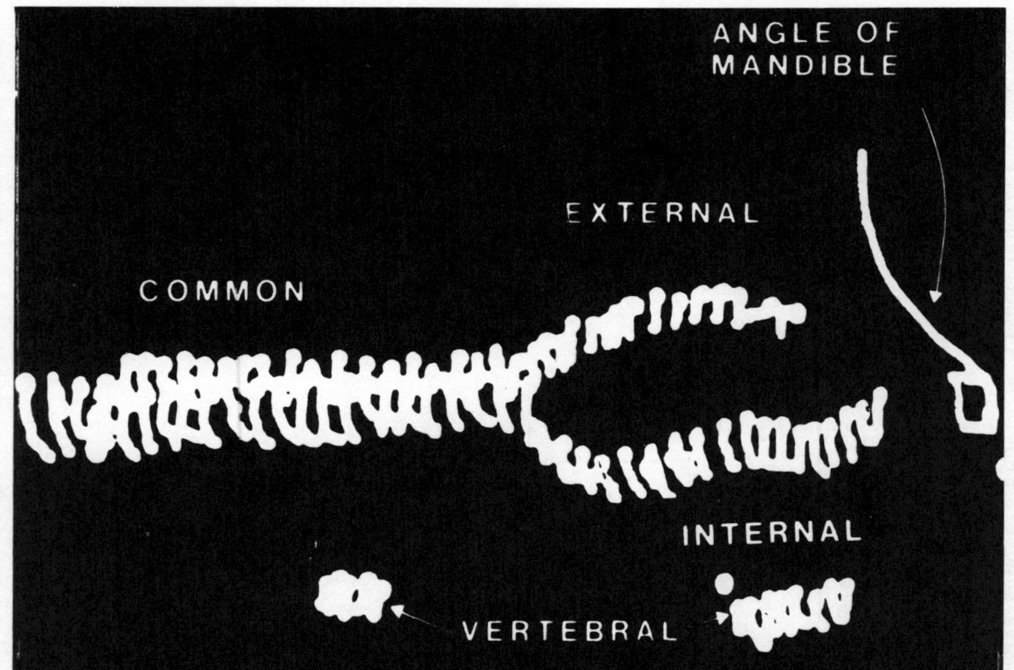

FIG. 7.12   Image of the carotid bifurcation and vertebral artery in the two positions posterior to the carotid projection. Angle of the jaw is indicated at the right by a square cursor. Vertebral artery Doppler image is seen in two positions, one posterior to the common carotid artery and in a vertebral artery where it is accessible to the sound beam in between the transverse processes of the cervical spine, and posterior to the internal carotid artery where it emerges to form the atlas loop.

7.2   TECHNIQUES OF CW DOPPLER EXAMINATION

7.2.1   Introduction

Table 7.1 identifies the major elements of our present non-invasive
cerebrovascular evaluation. A trained technician performs the rou-
tines and subroutines providing to the interpreter a worksheet with
the history and physical examination data, audio or video tapes with
voice notations, simultaneous spectral tracings and DOPSCAN images.
The complete examination by a skilled technician averages 40 minutes
including 10-15 minutes for the imaging. All Doppler examinations are
performed with the patient in the supine position and are illustrated
in this author's text (Spencer and Reid, 1981). The illustrations of
techniques used in this chapter demonstrate only recent approaches
in examination. The examination begins with the vertebral arteries
in the atlas loop, followed by the subclavian and then the ophthalmic
arteries in the posterior orbit. Imaging of the carotid bifurcations
is then performed. Subroutines, when indicated, include examination
of the retroclavicular common carotid and vertebral arteries, the
auxiliary and brachial arteries, testing for vertebral to subclavian
steal, temporal arteries, examination of the periorbital arteries,
arterial compression interventions, and hand-held examination of the
internal and external carotid arteries. "Down imaging" of the low
common carotid and origin of the vertebral arteries is also a subroutine.

7.2.2   Vertebral And Subclavian Artery Examinations

The vertebral artery signals may be detected in three diagnostically
significant locations:
- superior at the base of the skull (in the atlas loop) by the technique of Pourcelot (1976), chapter 5 and by Von Reutern et al
  1976a)
- inferior in the supraclavicular position by the technique of
  Brockenbrough (1981)
- along the spine between the cervical transverse processes (Fig.
  7.12) using Doppler imaging.

180

The qualities of the normal vertebral artery signal usually are modified along the vessel as it approaches the base of the skull. In the
low vertebral artery the normal signals possess a high pulsatility
index because of muscular branches supplied. At the base of the skull
the normal vertebral signals sound continuous because of elevated
diastolic flow producing a lower pulsatility index. The signal is
similar to the internal carotid artery as both directly supply the
brain (Fig. 7.13).

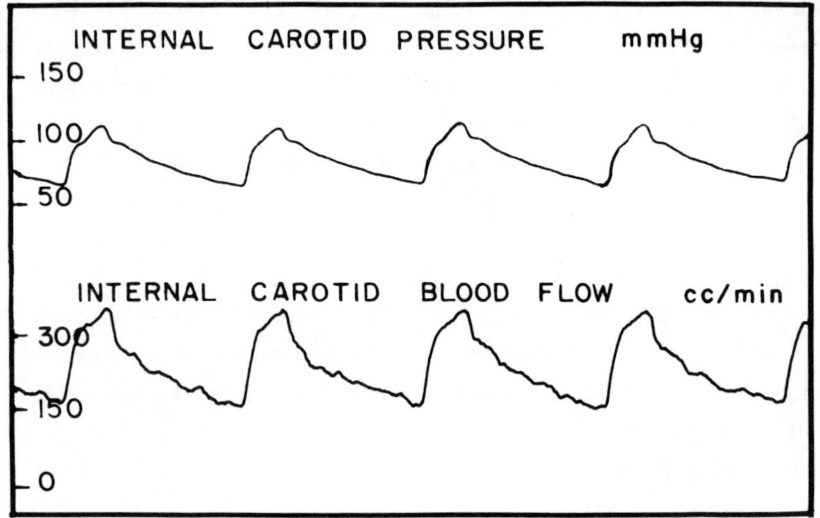

FIG. 7.13   Tracings from human internal carotid artery representing pressure pulse simultaneously with internal carotid
blood flow pulse recorded with electromagnetic flowmeter.
Similarities of the wave form between the two wave forms can
be seen. Exponential drop in pressure during diastole is accompanied by an exponential decrease in blood flow.
1 mmHg = 0.133 kPa.

Location of the brachial artery and its use in measuring systolic
arterial pressure is illustrated in Fig. 7.14. These examinations are
performed when there is suspicion of subclavian artery or innominate
obstruction such as when there are auscultatory pressure differences
between the two arms, when symptoms of basilar arterial insufficiency
occur or when there is a bruit in either supraclavicular region.
Fig. 7.15 illustrates the normal brachial artery signal in both the
resting and hyperemic state.

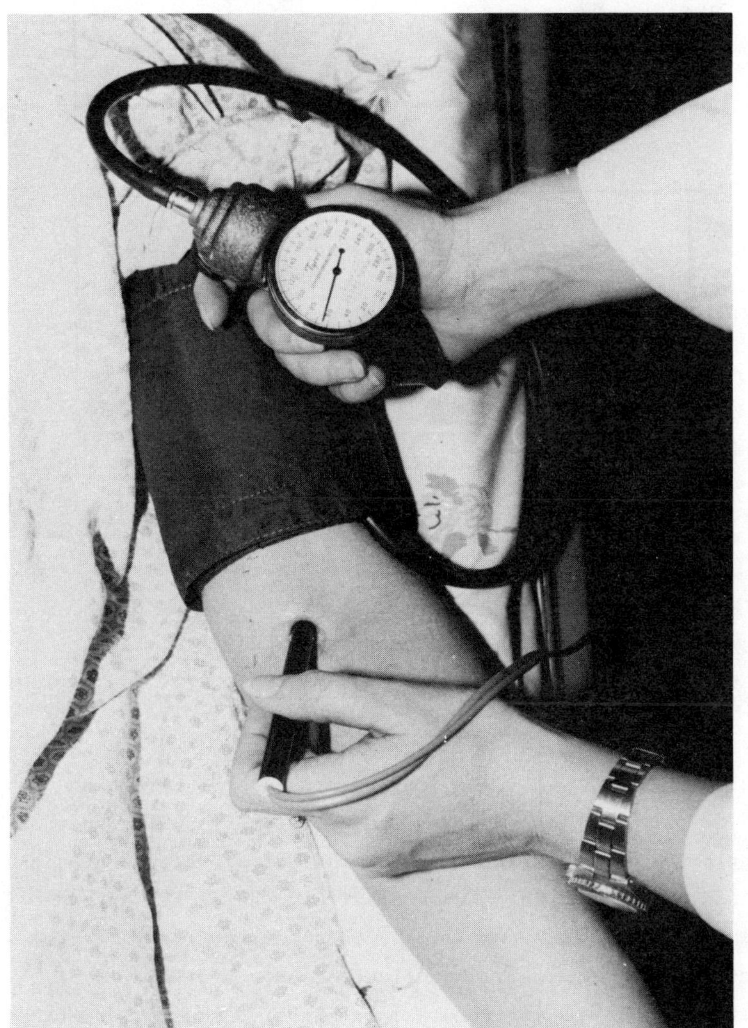

FIG. 7.14 Method of obtaining the peak systolic blood pressure in the brachial artery using standard sphygmomanometer and 10 MHz Doppler probe.

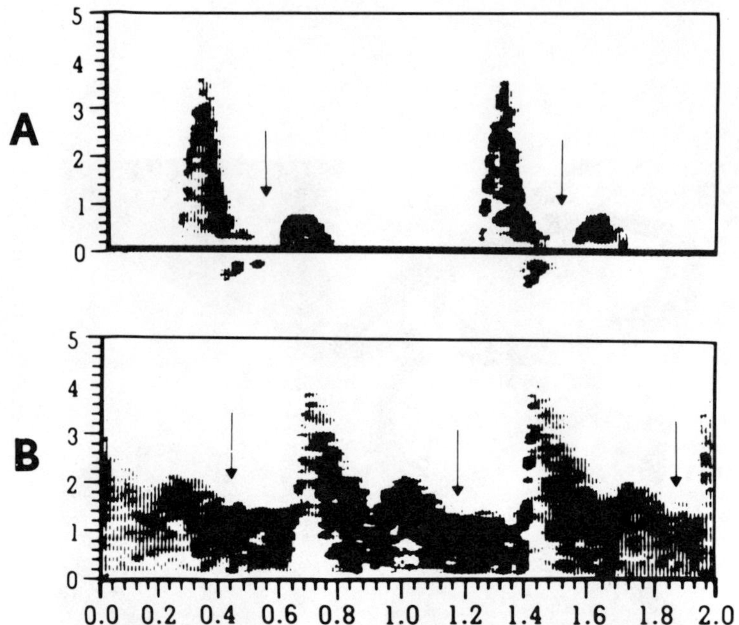

FIG. 7.15 The normal brachial artery signal,
Panel A and same artery signal during reactive
hyperemia, Panel B. Backflow during the dia-
crotic notch is seen in the normal resting arm.
During the hyperemia, the diastolic velocities
are elevated much more than systolic velocities.
Ordinates in kilohertz and abcissa in seconds.

If a bruit is found low in the neck, hand-held exploration or special
"down-imaging is employed. Down imaging consists of reversing the
angle of the probe as well as the Doppler polarity switch so that
the common carotid artery in its more proximal segments behind the
clavicle can be imaged. The vertebral artery, below its entrance into
the sixth lateral process, can also be detected and the vertebral artery stenosis at this level identified (Spencer and Reid, 1981).

The vertebral to subclavian steal examination should be performed
for the same indications as that for the upper extremity subroutines
or when the atlas loop vertebral signal is abnormal. To test for a
vertebral to subclavian artery steal it is preferable to utilize the
vertebral artery atlas loop signal, because of its greater reliability and greater brain significance when a positive test is present.

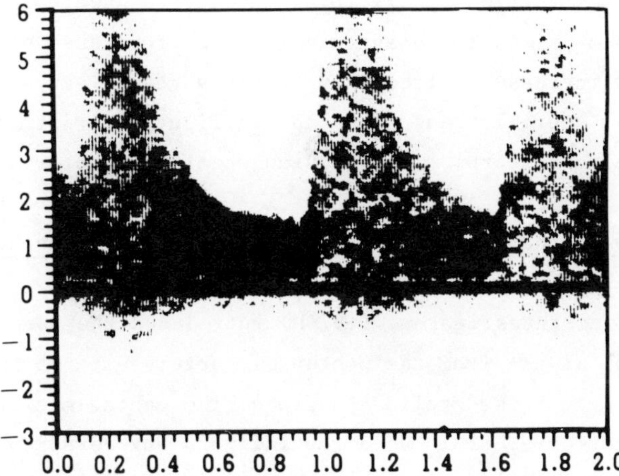

FIG. 7.16   Spectral displays of abnormal vertebral artery signals characteristic of vertebral to subclavian steal. The top panel illustrates biphasic flow in the vertebral artery and at the atlas loop during control. The bottom panel is from the same position following deflation of the arm cuff and during reactive hyperemia response in the arm. The augmentation of the signal in the upper direction proves that the upward directed spectrum is in the direction of the brachial artery and that biphasic steal is converted to continuous steal by reactive hyperemia in the homolateral arm. Ordinates in kHz and abscissa in seconds.

If a steal is detected its clinical significance may be assessed by
repeating the reactive hyperemia response in the supine position but
upon deflating the arm cuff put the patient in the sitting position.
If this maneuver produces or exacerbates symptoms of vertebro-basilar
insufficiency the clinical significance of the steal for the brain
is established and a more aggressive surgical approach may be considered. We suggest, for greatest response, the cuff be inflated for
3 minutes, and at the end of this period request the patient to perform a vigorous fist clenching just before release of the cuff.

The vertebral steal test as conceived in our laboratory by
Brockenbrough in 1976 depends on detecting an increase in the vertebral artery velocity caused by a sudden drop in subclavian artery
pressure. The pressure drop is produced downstream to subclavian or
innominate obstruction by eliciting a reactive hyperemia in the arm
on the suspect side. The diagnosis is made if release of the cuff
produces an increase in frequency in the vertebral artery signal
(Fig. 7.16). Pourcelot has given an excellent description of the
various analog patterns of subclavian steal in chapter 5.

7.2.3   The Ophthalmic Artery And Its Periorbital Branches

Spencer and co-investigators (1974b) introduced the use of the posterior orbital signal from the ophthalmic artery using 5 MHz probes and
focussing through the eyelid and globe. The ophthalmic artery signal
in the posterior orbit is used routinely in our cerebrovascular evaluation because, being closer to the Circle-of-Willis than its periorbital branches, it reflects the pressure gradient between intra and
extracranial arteries. The 5 MHz probe is placed over the eyelid, directing the ultrasonic beam through the eyeball at an angle of 10-15°
from the midline plane (Fig. 7.17). Angles greater than 20° should be
avoided because of distortion of directional and frequency characteristics. The patient is requested to direct their gaze towards the
opposite side in order to remove the lens of the eye from the path of
the sound beam. Several signals may be found from intraorbital branches of the ophthalmic artery but the strongest signals should be
sought and recorded. Qualities of the normal ophthalmic artery signal
are similar to those of the normal internal carotid artery but the

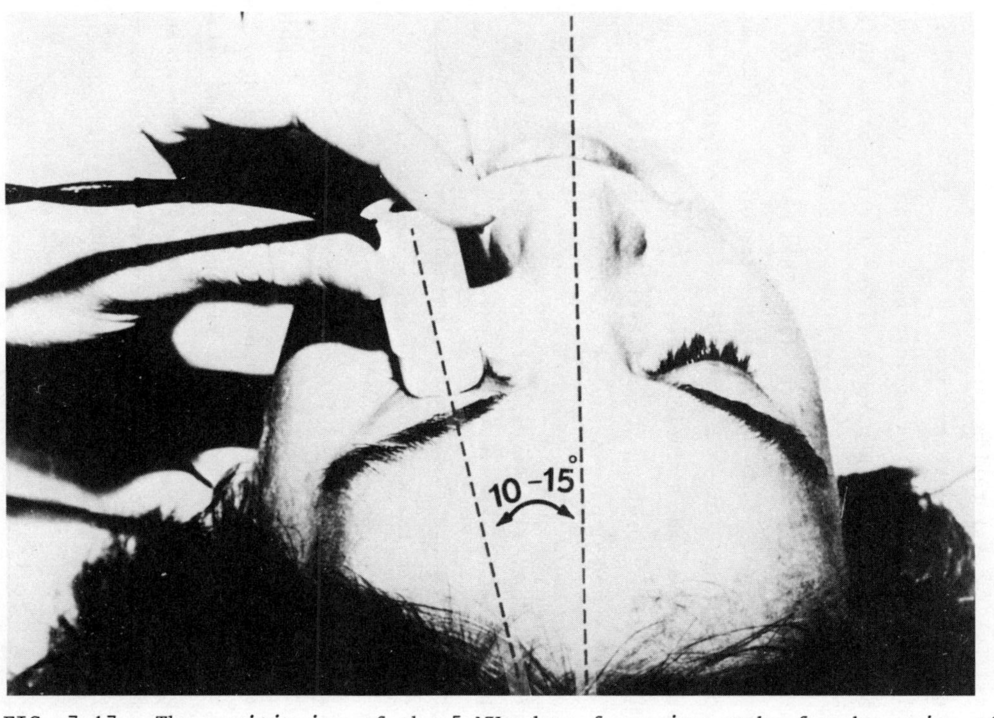

FIG. 7.17   The positioning of the 5 MHz deep focussing probe for detection of the ophthalmic artery signals behind the eyeball in the posterior orbit.

diastolic component may be diminished or increased depending on the percent of distribution to the scalp. A strong reversed high frequency signal (greater than 6 kHz) is diagnostic of significant obstruction of the internal carotid artery. A normally directed signal (towards the probe) may also be present in spite of carotid obstruction and skillful interpretation depends on other abnormalities such as low frequency and slow acceleration.

Periorbital collateralization effects have been recently revised by Brockenbrough (1981). Brockenbrough (1976) first applied shallow-focussing, 10 MHz probes to the superficial branches of the ophthalmic arteries found around the orbital rim. This method utilizes the signal changes evoked by compression of the temporal, mandibular or common carotid arteries. It is a test for carotid collateral supply around obstructions of the carotid arteries. Great skill is required in performing and interpreting the collateral evaluation tests. Barnes and Wilson (1975) have produced an excellent manual to assist the beginner.

The periorbital arterial signals are best obtained with the shallow-focussing, 10 MHz probe. The supraorbital artery is found as it courses through the supraorbital notch, palpable on the superior orbital rim just beneath the eyebrow. The probe beam should be directed well under the rim taking care to avoid the palpebral artery within the lid itself. The direction of flow is normally from within the orbit to the forehead subcutaneous tissues but the indicated Doppler direction is unreliable because of changes in the direction of the artery as it passes around the orbital rim and also because of reflections of the sound beam from underlying bone. The frontal and angular artery signals are found at the inner canthus of the eye but the direction of these signals is also unreliable.

7.2.4   Carotid Compression Techniques

Selective common carotid artery compression (CCC) is performed as a subroutine when clarification of collateral pathways is needed. It is performed when there is an apparent discrepancy between the periorbital or posterior orbital artery signals and the carotid artery Doppler signals. For example, when severe carotid artery stenosis is

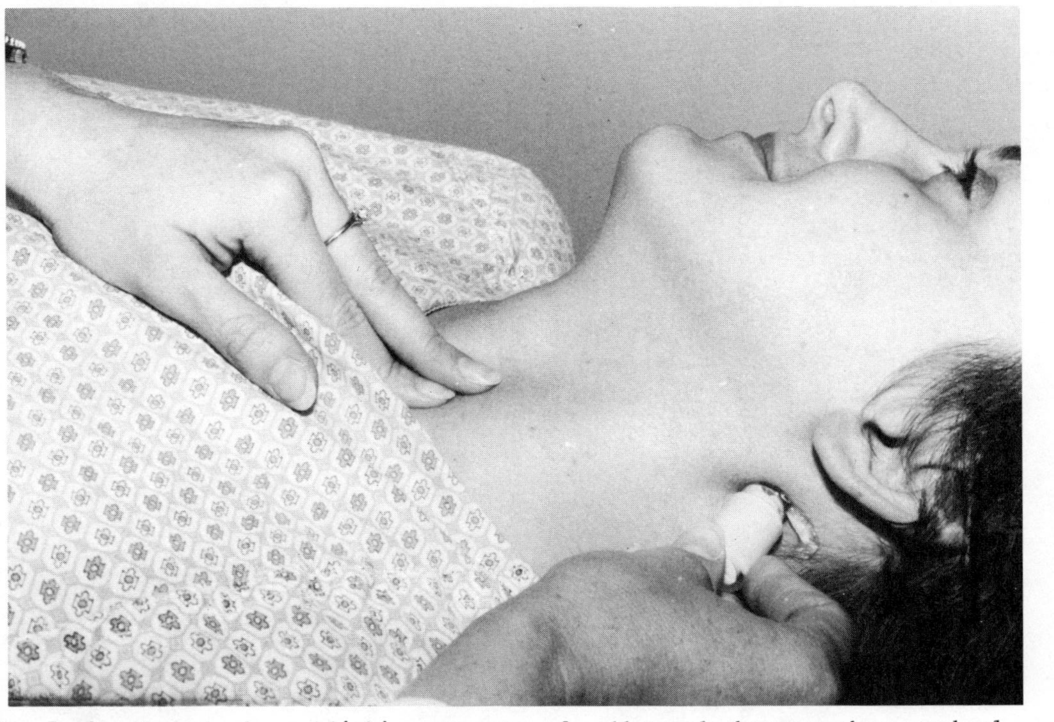

FIG. 7.18   Method of establishing presence of collaterals between the vertebral artery and the common carotid artery. 5 MHz lens focussing probe is placed behind the mastoid process to detect the vertebral signal in the atlas loop. While recording the Doppler, the common carotid artery may be compressed low in the neck, below the bifurcation. Augmentation of the vertebral artery signal during this maneuver establishes an intracranial posterior-to-anterior connection.

indicated by carotid bifurcation signals in the neck by either DOPSCAN or hand-held techniques and yet the orbital signals appear normal, contralateral CCC is of great value to establish the presence of intracranial collaterals across the anterior communicating artery. In this case contralateral compression will diminish the antegrade ophthalmic artery signal. The compression may be carried out while interrogating either the posterior orbital artery signal, the periorbital artery signal or the vertebral artery signal (Fig. 7.18).

For safety, CCC should be performed at a position low in the neck just above the clavicle and for only 2-3 heart beats. It should be performed slowly, without quick movements, to minimize stimulation of baroreceptors. The ear pulse should be monitored to assure that compression is complete and to reduce the amount of manipulation of the carotid artery. In thousands of compressions performed in our laboratories by technicians, nurses and physicians, no complications have been observed other than occasional bradycardia. The bad reputation of past carotid artery compressions undoubtedly arose from improper techniques, such as compression at the bifurcation (where the pulse is most easily found) and continuing the compression to elicit symptoms and signs of cerebral ischemia. Carotid artery compression may be inadvisable if cervical carotid Doppler examination identifies a low bifurcation. We recommend this maneuver be carried out as a subroutine at the end of the examination when the level of the bifurcation has been identified and when the additional information is needed concerning collaterals.

7.2.5   Hand-held Carotid Probing

If Doppler imaging of the carotid bifurcation leaves uncertainty concerning patency of the internal carotid artery, hand-held probing is useful to complete the search. The common carotid artery should be detected low in the neck just above the clavicle. The internal and external carotid artery signals are most accurately identified high in the neck 2-3 cm below the mandible. One should first approach the internal carotid artery which usually has a stronger signal than the external carotid artery. It is more posteriorly placed and the probe

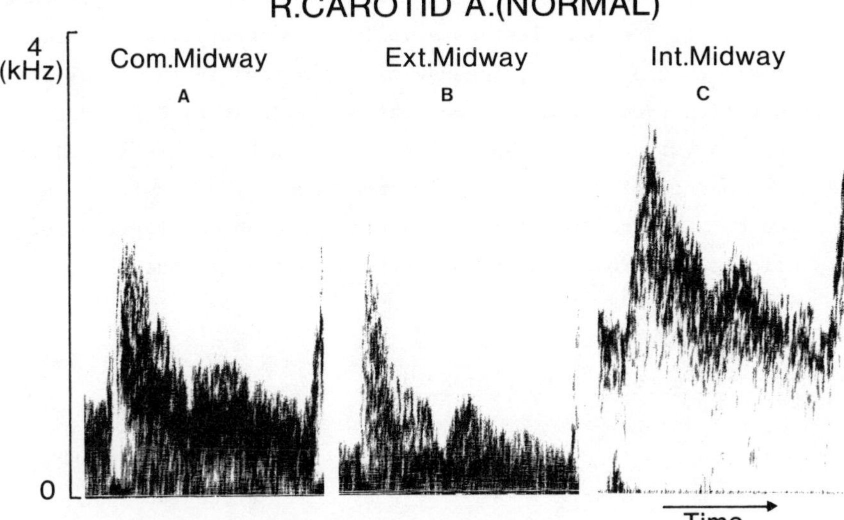

FIG. 7.19   Audio presentations of typical spectra from the normal common, external and internal carotid arteries.

beam should be directed slightly posterior with a superior angle 30-
45° from the body axis. The probe should contact the skin at a point
approximately 3 cm below the angle of the jaw and pointed in the direction of the mastoid process. The typical qualities of the internal
carotid signal (Fig. 7.19C) may be found by slowly moving the probe in
small searching angles until the best signal is detected. After finding and recording the internal carotid artery signal the probe is
angled in a medial and anterior direction while listening for the

typical qualities of the external carotid artery (Fig. 7.19B). Movement back and forth between these two signals will convince the operator of their spatial and physiological differences.

Excellent hand-held Doppler techniques have been advanced by Itti et al (1978), by Von Reutern et al (1976b) and by Mol in chapter 6. Hand-held techniques may have limitations in the hands of other than the highly skilled technician, especially when separating less than 90% stenosis from occlusion, because of variations in the level of the bifurcation from patient to patient and because of difficulties in locating a very narrow stenotic channel. Von Reutern et al (1976b) have developed to a high skill their technique of performing a hand-held "run through" recording the common to internal analog recordings while sliding the probe beam through the stenotic lesions (Fig. 7.8).

7.3   DIAGNOSTIC HEMODYNAMICS

7.3.1   Normal Carotid Artery Dynamics

The nominal values for various features of the blood flow in the normal internal carotid artery has been previously presented (Spencer and Reid, 1981). Concentration of energy near the maximum velocity (Fig. 7.19) occurs because of a flat and laminar profile in the normal flow streams. The contour of the maximum velocity edge of the spectrum in Fig. 7.19 as well as the mean flow of Fig. 7.13 is resistive in shape, as explained in chapter 6, closely following the arterial pressure pulse. The flow pulse contour is dominated by low peripheral resistance of the brain. Pourcelot has discussed in chapter 5 how the higher peripheral resistance of the external carotid artery increases the pulsatility index in the external carotid artery and how an increase in pulsatility in the common carotid artery provides a clue to internal carotid stenosis. Reneman and Hoeks in chapter 4 and Gosling in chapter 8 illustrate well the manifestation of the difference in compliant flow pulse in the common, external and internal carotid arteries.

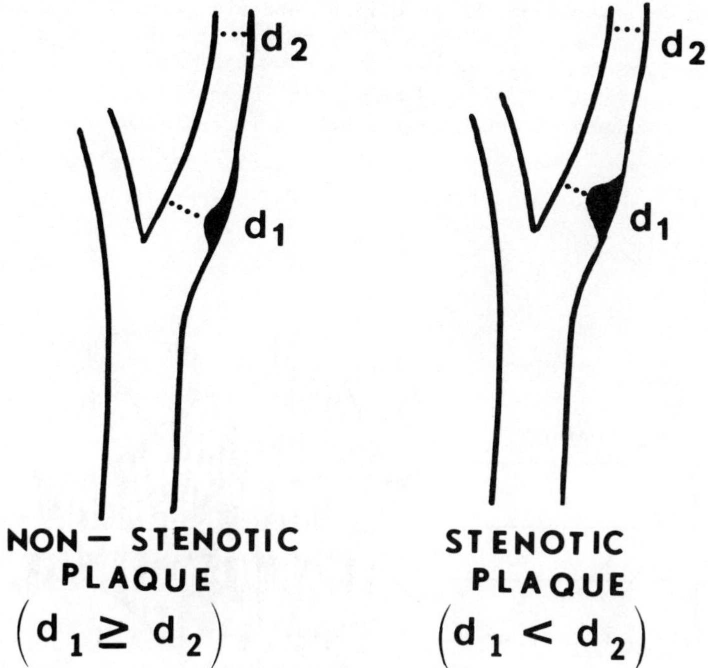

FIG. 7.20   Non-stenotic plaque is defined as the downstream diameter less than or equal to the internal carotid artery bulb diameter. Functional internal carotid stenosis is defined as the diameter in the stenosis less than the downstream diameter.

7.3.2   <u>Carotid Artery Abnormalities</u>

7.3.2.1   Definitions

"Obstruction" is defined as including any condition of stenosis or occlusion. "Occlusion" means complete obliteration of the lumen. "Flow" is a volumetric term expressed as volume per unit time, e.g., ml/min. "Velocity" is the rate of change of position, e.g., cm/sec. "Stenosis" is defined, for the purposes of Doppler diagnosis, in Fig. 7.20. If the plaque is not producing stenosis we apply the term non-stenotic plaque. In other words stenosis is defined as a cross-section less than the downstream normal cross-section (usually beginning 4 cm above the bifurcation). Both stenotic and non-stenotic plaques are detectable at the carotid bifurcation with DOPSCAN and SONACOLOR. In addition, increased intracranial artery and carotid siphon lesions and increased brain vascular resistance can be diagnosed. Mol in chapter 6 demonstrates how an increase in brain resistance (also siphon stenosis) increases the relative compliant velocity component and the pulsatility index in the internal carotid artery.

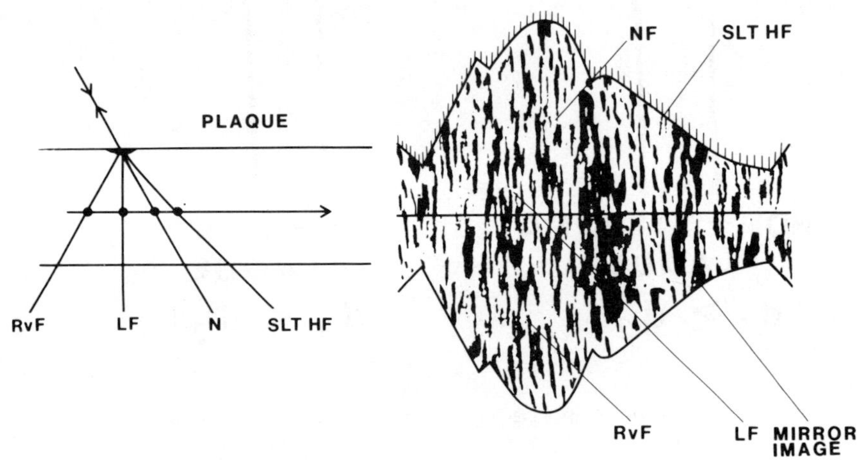

FIG. 7.21   Explanation of how spectral broadening may be produced without turbulence by scattering of the sound beam caused by irregular plaque material in the artery wall. The reversed spectrum is symmetrical in contour to the forward one. RvF = reversed frequency; LF = low frequency; NF = normal frequency; SLTHF = slightly high frequency.

7.3.2.2 Non-stenotic Plaque

Since the detected ultrasound beam must traverse the arterial wall twice, calcium deposits and roughening of the luminal surface easily produces recognizable spectral broadening (Fig. 7.21) and other abnormalities in the Doppler signal (Spencer and Reid, 1979). Recent in vitro studies of excised iliac arteries has confirmed that calcium deposits in atherosclerotic plaques can produce both asonic gaps in the DOPSCAN image as well as inverted signals (Fig. 7.22; Spencer and Reid, 1981). In a series of zero-radiography studies on 31 carorid endarterectomy specimens, 30 disclosed dense amorphous calcium deposits. The single specimen, without calcium, was determined to represent fibromuscular dysplasia. Table 7.4 summarizes the CW Doppler diagnostic features of non-stenotic plaques. An asonic gap, an inverted signal, or signals of turbulence in the absence of local frequency elevations are singularly diagnostic of a non-stenotic plaque. Also, the diagnosis of "loss of the bulb" due to non-stenotic plaquing can be made when the normal progression of frequency increase along the first four centimeters of the internal carotid artery is replaced by a constant frequency along this segment.

TABLE 7.4 : The CW Doppler diagnostic features of non-stenotic plaques

SUFFICIENT TO MAKE DIAGNOSIS	NEED THREE OR MORE QUALITIES
1. Non-sounding segment - total scattering of sound beam by dense calcium deposits	1. Coarseness - (Spectral widening)
	2. Weak signals
2. Inverted signals - scattering effect of calcium producing change of sound beam direction	3. Biphasic - changing phase during heart cycle.
	4. Systolic - normal diastolic component appears.
3. Fluttering - turbulence due to wall roughness	All qualities caused by beam deflection and scattering

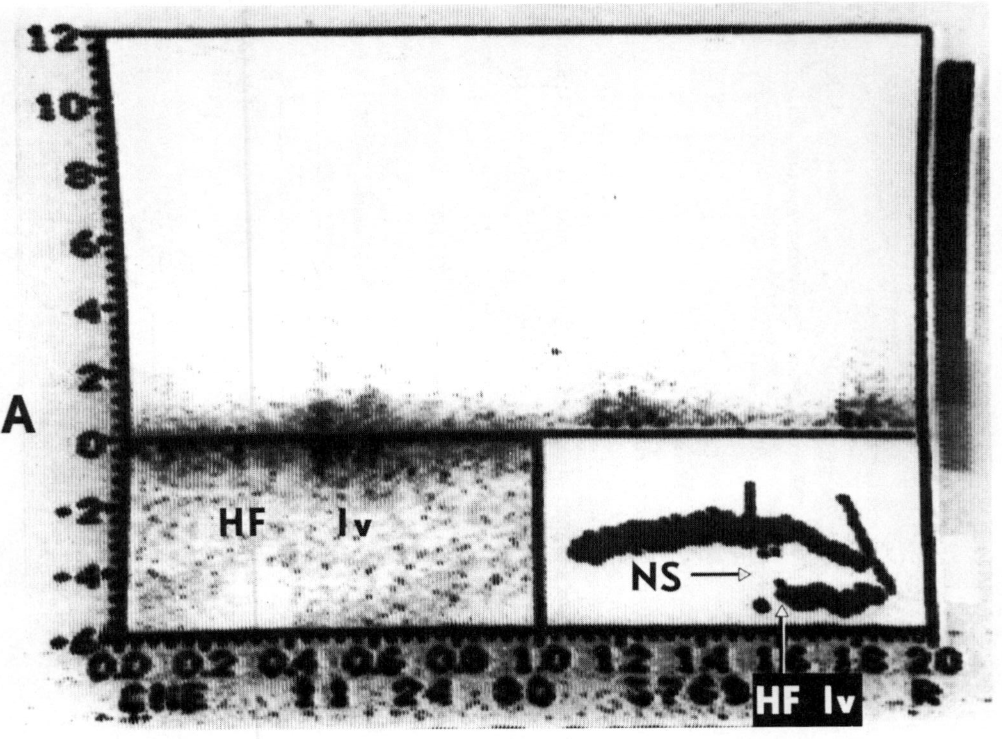

FIG. 7.22A   Illustration of the effect of dense calcified plaquing on
continuous wave Doppler signal, producing an asonic gap. This does not
usually impede the detection of stenosis high frequencies, here seen
inverted (Iv) by the bending of the sound beam at the end of the calcium deposit.

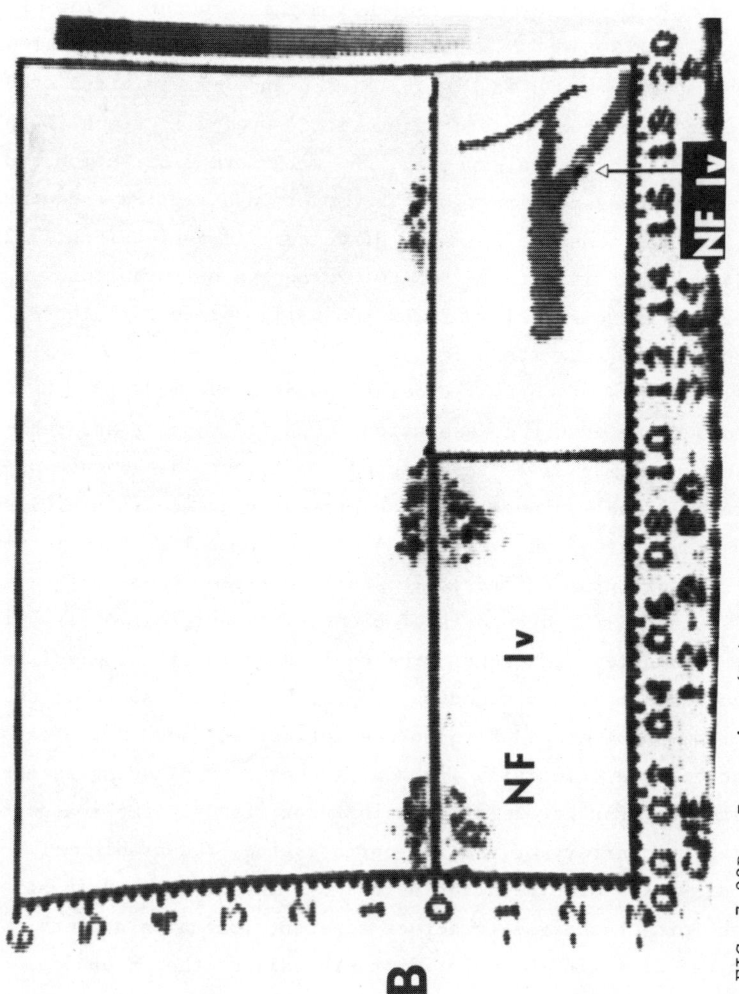

FIG. 7.22B Inversion (Iv) of the energy primarily by a plaquing on the origin of the internal carotid artery. Although the energy is returning from both directional senses, most of the energy is in the inverted direction. NF = normal frequency.

7.3.2.3 Spectral Manifestations Of Turbulence

Localized turbulence is produced when wall roughening of stenosis is sufficient to break-up laminar flow streams. This may be caused by non-stenotic plaques when abrupt widening of the channel occurs. Within a crater the eddies of turbulence may also occur. Using directional CW Doppler SONACOLOR, turbulence is registered in several different ways as diagrammed in Figs. 7.5, 7.6 and 7.7. _In vitro_ studies of turbulence downstream to stenosis, with a dual-directional Doppler velocimeter, discloses dual spectra, one with normal direction and the other with reversed direction. The reversed direction frequencies are usually lower than the normally directed frequencies (Fig. 7.23). Figs. 7.24 and 7.25 illustrate SONACOLOR spectra of turbulence caused by non-stenotic plaques. Figs. 7.26B and C illustrate turbulence spectra downstream to a stenotic plaque.

A true diagnosis of intimal ulceration cannot be made from Doppler ultrasound because even the resolution of pulsed Doppler and real-time pulsed echo imaging is insufficient to detect the absence of the intimal layer. Equally invalid is the present practice in radiology to diagnose ulceration when cratering of the luminal surface is seen on the X-ray angiogram. Microscopic studies of the lumen surface is the final standard for diagnosis of ulceration. Surgical observation is often valid unless adherent thrombus is seen in histological sectioning but this may be misleading. A presumptive diagnosis of embolization from an ulcerated site on the intimal surface of the carotid artery can be made clinically if local stenosis or plaquing is associated with TIA's or stroke. Amaurosis fugax strongly implicates the internal carotid artery of the appropriate side. The problem of non-invasive diagnosis of ulceration points out the great need in stroke prevention for a technique to detect adherent lumenal platelets, fibrin or red clots in the carotid artery wall so that steps can be taken to prevent their embolization to the brain. Real-time pulsed echo imaging can occasionally do this but its accuracy requires validation for routine use.

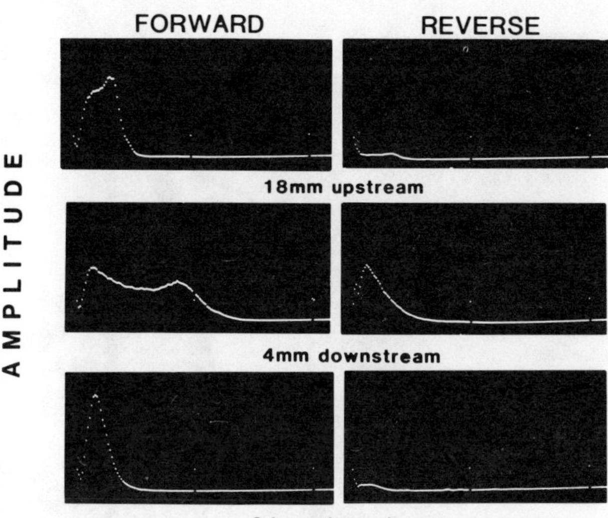

FIG. 7.23 Doppler velocity spectra obtained from <u>in vitro</u> stenosis studies. Both forward and reverse directions are indicated. Upstream to the stenosis a slight reversed spectrum is represented because of the limitation of the Nippa circuit to completely separate forward from reversed signals. In the middle two panels, 4 mm downstream to the stenosis, the forward going spectrum is broadened from 0 to 2.5 kHz. Increased energy in the reversed direction is apparent but lower frequencies are present. The lower two panels illustrate how the flow returns to a laminar pattern without spectral broadening and without reverse flow signals.

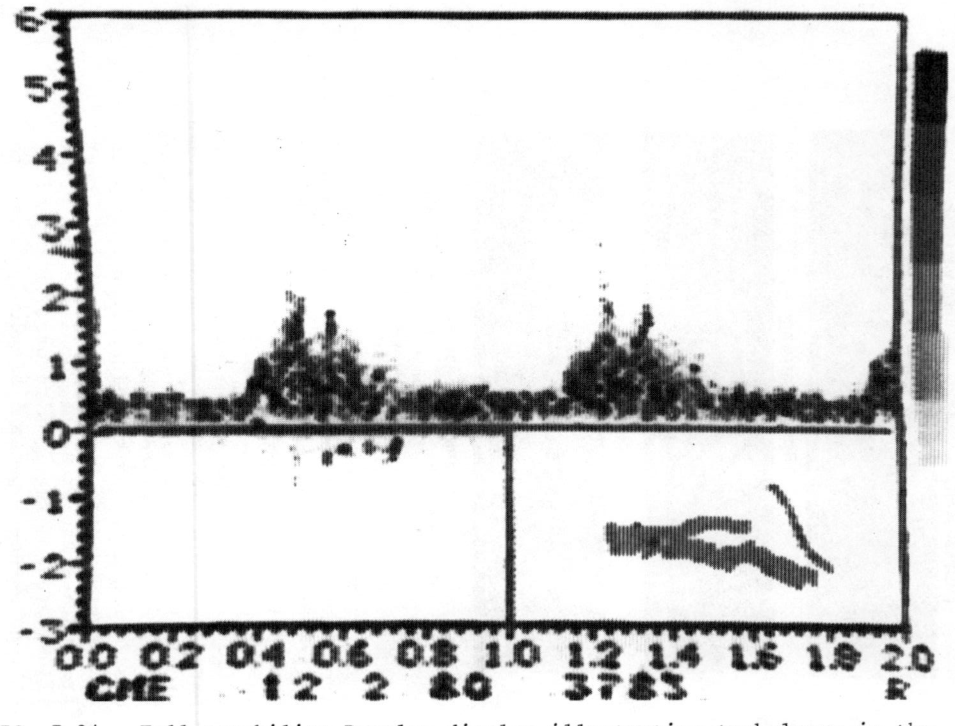

FIG. 7.24   Full capability Doppler display illustrating turbulence in the common carotid artery, just below the bifurcation. Irregularity, in the upper edge of the spectrum during peak systole, sometimes "whipping" in quality, provides a repetitive pattern from beat to beat during systole. This is the earliest form of turbulence and is not usually sufficient to produce a bruit.

FIG. 7.25   Later stage of turbulence than seen in Fig. 7.24. Here irregularity of the upper edge of the spectrum persists from systole down into diastole. This pattern indicates sufficient energy to produce a grade I bruit during systole.

7.3.2.4 Carotid Artery Stenosis

The Doppler manifestations of arterial stenosis include primary, secondary and tertiary hemodynamic abnormalities.

Hemodynamic Effects Of Arterial Obstruction

Primary: Greater velocity within stenosis
- Usually elevated above normal values
- Always greater within segmented stenosis than in downstream channel.

Secondary: Downstream effects
- Usually turbulence from segmented stenosis or
- Lower than normal velocities and flow in the downstream channel and immediate branches when stenosis is severe.

Tertiary: Collateralization effects
- Increased flow and velocity in adjacent branches or alternate contralateral arteries. This can also lead to turbulence in the collateral channels
- Diminished flow and velocity in interconnecting arteries
- Reversed flow in immediate branches of stenotic artery. Reversal may be transitory (to and fro flow within the heart cycle when the pressure gradient is low and pressure pulses are arriving from both directions).

The Doppler characteristic of increased flow stream velocities is an increase in frequencies manifest on the frequency spectrum. Decreased velocities are represented by decreased frequencies and reversed flow direction is manifest by backflow at low to high velocities depending on the magnitude of the reversed pressure gradient.

The primary feature, increase in Doppler frequency, follows directly the increase in blood velocity which occurs as the blood streams accelerate into the stenotic segment (Fig. 7.26A). Secondary to increased velocity within the narrowed segment is the turbulence

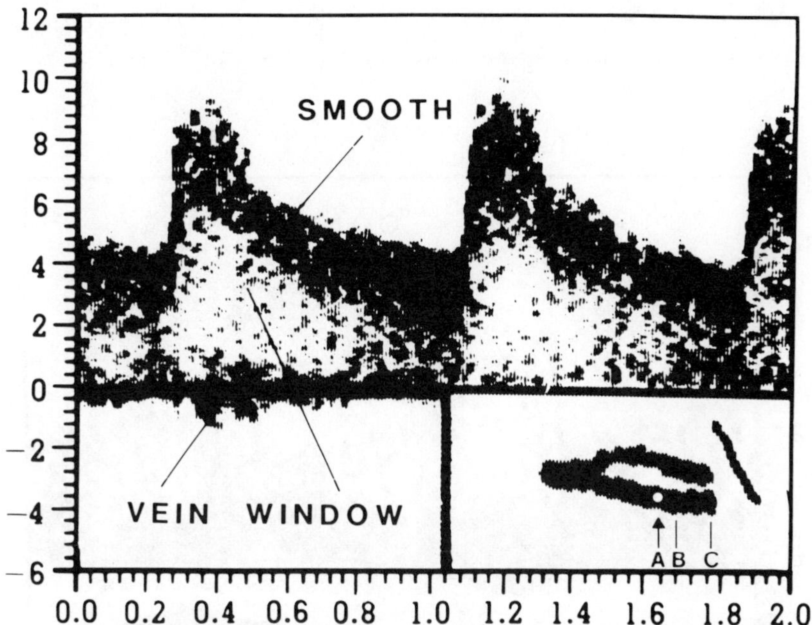

FIG. 7.26A   High frequencies representing primary feature of
high velocity within a stenosis of the internal carotid
artery at position A. Most of the sampled blood velocities
are present as laminar flow within the stenosis and evidenced
on the spectrum by a smooth concentration of energy near
the upper edge and a paucity of the energy below this producing a "window". Superimposed vein signal produces some
low frequency energy in the reversed direction below the
zero frequency in the minus direction.

that usually occurs immediately downstream as well as decreased velocities due to the restriction of volume flow. Figs. 7.26A, B, C and
D illustrate how DOPSCAN with SONACOLOR can detect and localize the
primary, secondary and tertiary features of carotid artery stenosis.
Tertiary features are caused by alterations in direction and velocity
of blood flow within the collateral arteries. The collateral effects
in carotid artery obstructions result from decreased intracranial arterial pressure downstream to the stenosis. With the combined advantages of CW DOPSCAN imaging all the primary, secondary and tertiary

features of carotid stenosis can be accurately diagnosed with the
severity of turbulence determined with great accuracy (Spencer et al,
1974a; Spencer and Arts, 1981). With the aid of new bi-directional
spectral analyzers such as SONACOLOR improved recognition of these
features will lead to greater speed and accuracy of diagnosis.

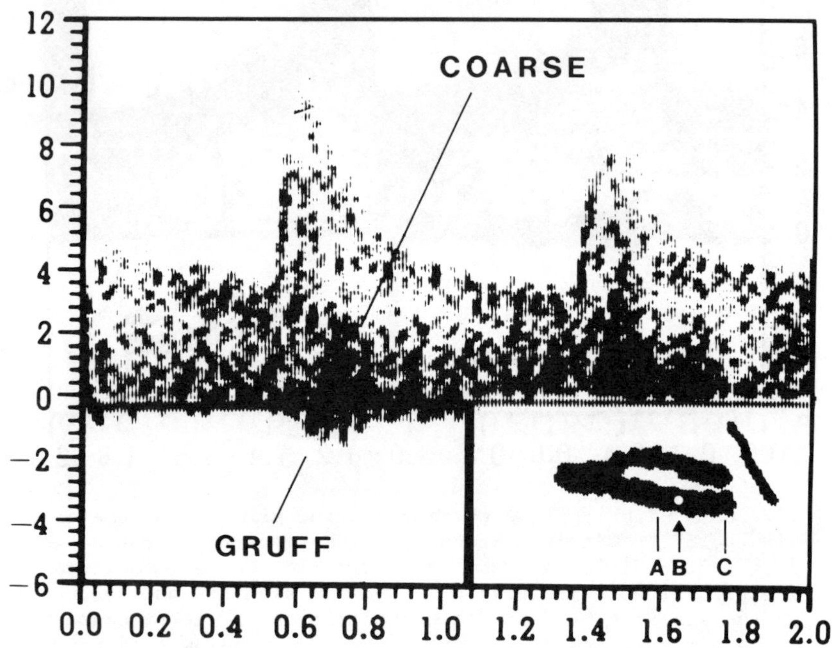

FIG. 7.26B   Downstream to the stenosis position (B) where
   high velocities of the emerging jet are still represented.
   The spectrum, however, displays secondary features of
   gruff qualities of the bruit evidenced by a concentration
   of energy during systole on either side of the base line.
   These frequencies are generally less than 1 kHz. In addition, spectral broadening produces a coarse sounding quality to the ear, is evidenced by a loss of concentration
   of energy near the maximum edge of the spectrum.

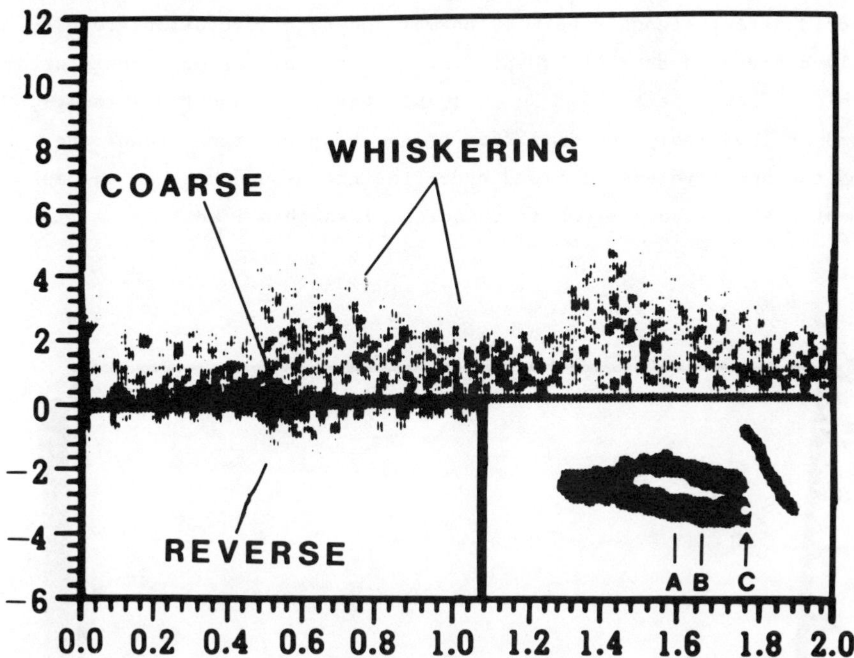

FIG. 7.26C   Considerably further downstream position (C),
the velocities are diminished, turbulence is still evidenced
by coarseness of spectral broadening and whiskering. Laminar
flow has not yet returned at this position.

The high frequencies within the stenotic segment are proportional to
the degree of stenosis (Spencer and Reid, 1979) (Fig. 7.27). The
degree of elevation in frequency, however, is subject to the richness
of collateral circulation (Spencer and Reid, 1981), the magnitude of
the arterial pressure as well as the viscosity and density of the
blood (Spencer and Arts, 1981). Comparisons of maximum Doppler frequencies and arteriographic diameters suggest DOPSCAN is as accurate
as X-ray angiography (Spencer and Reid, 1981) in estimating the degree of stenosis but proof must await a better standard of reference.
With the use of spectral analyzers, we can probably improve the Doppler estimate of percentage of stenosis. Comparing the Doppler frequency downstream or upstream to the frequency within the stenosis
we can calculate the percentage of change in velocity and thus the
percentage difference in cross-sectional area between the two

measurement points (Spencer and Reid, 1979). The grades of internal
carotid artery stenosis between normal and total occlusion are use-
fully designated as I to V depending upon the effect on carotid artery blood flow (Spencer and Reid, 1979). Grades IV and V are recognized by invariably low amplitude of the Doppler signal resulting
from the small volume of blood under the ultrasound beam in accompa-
niment with low downstream frequencies, less than 1 kHz.

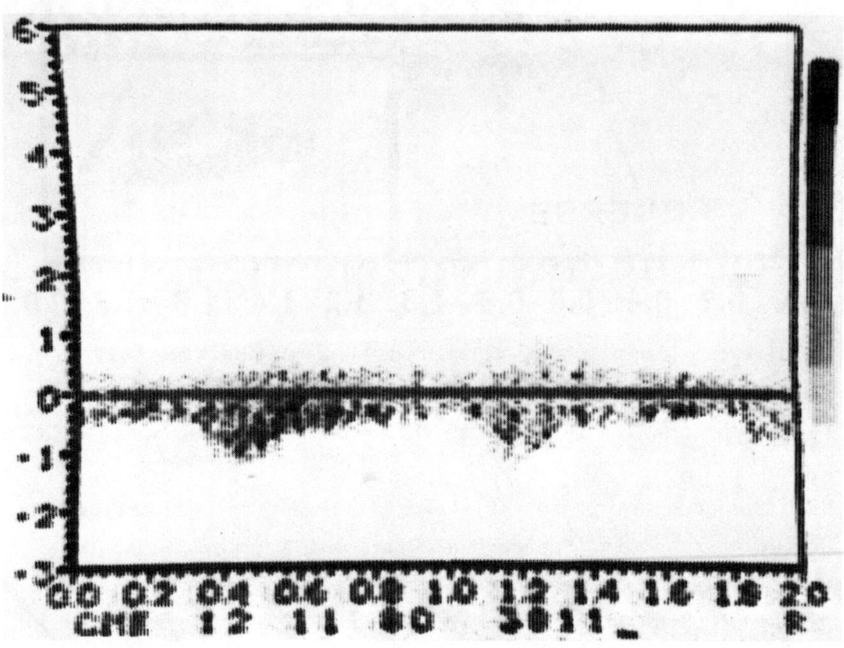

FIG. 7.26D   The posterior orbital ophthalmic signal in the
   same patient on the side of the internal carotid stenosis,
   illustrating tertiary features of carotid obstruction including slow acceleration and low frequencies. The flow signal
   is recorded in the normal anterior direction indicating the
   normal pressure gradient is reduced but not reversed.

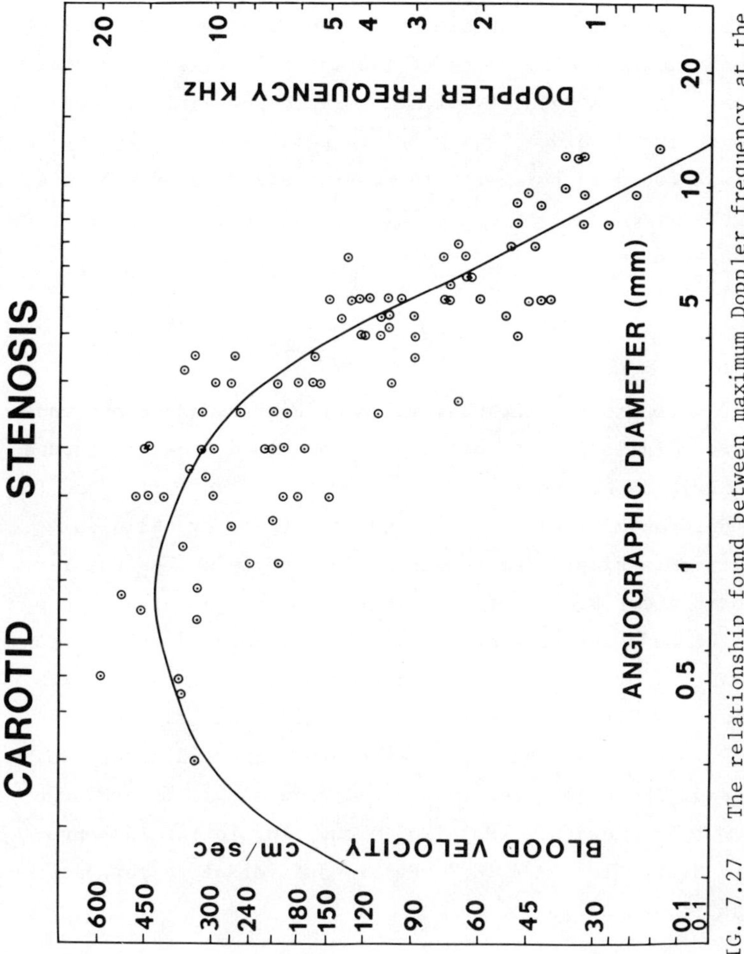

FIG. 7.27 The relationship found between maximum Doppler frequency at the origin of the internal carotid artery and the angiographic diameter of the artery at that position. This data is taken from 96 internal carotid arteries, angiogrammed after Doppler analysis. Maximum Doppler frequency of 5 kHz represents approximately 3.5 mm diameter of the internal carotid stenosis. This data was collected before the availability of on-line spectral analysis.

206

The pressure drop across an arterial stenosis can be calculated by application of the Bernoulli principle in chapter 3:

$$\Delta p_s = \tfrac{1}{2}\,\rho v^2$$

This principle was tested in experimental carotid artery stenosis of the sheep. Δp was measured by means of two lateral pressure points along the artery and the velocity within a region of variable constriction was measured between the pressure points. The following relationship was found to hold with increasing severity of stenosis, up to the highest frequency of Grade III stenosis and so long as turbulence was detectable:

$$\Delta p_s\,(\text{mmHg}) = 0.3\,f^2_{max}$$

The factor 0.3 accounts for the conversion of units and various underlying assumptions including the density of the blood, the ultrasound frequency (5 MHz), the probe angle (60° from the vessel axis) and the relationship between maximum velocity and mean velocity. Clinical experience with this simplified relationship has yielded useful results which have never been ambiguous. For example, the highest Doppler frequency we have encountered at an internal carotid artery stenosis was 20 kHz. The calculated pressure drop was 120 mmHg (16 kPa) but the patient was hypertensive with a systolic brachial pressure of 205 mmHg (27.3 kPa) and the opposite internal carotid artery was totally occluded. The calculated systolic Circle of Willis pressure was realistic at 65 mmHg (8.7 kPa) (205 Minus 120) in this asymptomatic patient. Sixty-five mmHg (8.7 kPa) is sufficient to perfuse the brain asymptomatically.

7.3.2.5 Carotid Collateral Circulation

Doppler tertiary evidences of collateral compensation for internal carotid artery obstruction can be found in the external carotid artery at the bifurcation, in the ophthalmic artery in the posterior orbit, in the branches of the ophthalmic artery, in the temporal and in the vertebral arteries. If there is no carotid artery stenosis the ophthalmic artery does not function as a collateral, ophthalmic artery flow is anterior in direction and compression of the ipsilateral temporal or mandibular artery will either produce no change in the periorbital artery signals or produce an augmentation of the signal in the normal direction. The diagnostic features sought as evidence of internal carotid artery collateralization are a decrease, obliteration or change in direction of the flow signals. This occurs because the pressure in the anterior Circle of Willis is diminished reversing the flow so that the external carotid artery functions as a collateral, supplying the otherwise normal territory of the ophthalmic artery and the internal carotid artery when the pressure drop is great. When this occurs, there is an obvious increase in velocity noted in the external carotid artery. Other internal carotid collateral effects are discussed in the chapter by Mol.

Frequently no or minimal collaterals will be detected from the homolateral external carotid artery even when the internal carotid artery is severely stenosed or occluded. If a normally directed ophthalmic artery signal is arising from an anterior communicating artery collateral, compression of the opposite common carotid artery will diminish the ophthalmic artery signal. In this case homolateral compression of the common carotid artery will either augment or not affect the signal. Also, if compression of the other common carotid artery causes augmentation of the vertebral artery signal in the atlas loop the presence of collateral channels from the posterior circulation is established. Table 7.5 provides a grading system for collateral circulation in areas perfused by the external, internal and vertebral arteries.

7.3.2.6 External And Common Carotid Obstructions

The diagnosis of external carotid artery stenosis (Figs. 7.28A and B) is important in the patient with an asymptomatic bruit because if no additional stenosis is found in the internal carotid artery this diagnosis somewhat reduces concern. If both external and internal carotid artery stenosis exists, judgement of the collateral evidences

TABLE 7.5 : Grading collaterals around internal carotid artery stenosis

---

Doppler evidences of collaterals around internal carotid obstruction:

A. <u>Homolateral ophthalmic artery</u>

Reversed direction with HF	= Grade 4
Reversed direction with LF	= Grade 2
Bidirection or dual direction with LF	= Grade 1
Normal direction with NF	= Grade 0

B. <u>Contralateral common carotid artery compression</u>

Changes ND ophthalmic to Rv	= Grade 4
Obliterates ophthalmic	= Grade 3
Decreases ophthalmic frequency 1/2	= Grade 2
Decreases frequency less than 1/2	= Grade 1
No change in ophthalmic or periorbital	= Grade 0

C. <u>Basilar artery</u>

Vertebral artery signal St & HF greater than 4 & CCC Augment	= Grade 2
Vertebral artery signal St, f=2-4 kHz	= Grade 2
Both signals strong with HF but does not augment	= Grade 4
Less than 2 kHz does augment	= Grade 1
No augmentation	= Grade 0

HF = high frequency; LF = low frequency; NF = normal frequency; ND = normal direction; Rv = reversed direction; St = strong signal

must be compensated. For example, simultaneous external and internal carotid artery obstructions like common carotid artery stenosis reduces the velocities in the ophthalmic artery but may not change its direction (Fig. 7.26D). This occurs because the pressure in both the Circle of Willis and the external carotid artery branches is diminished but the pressure gradient remains in the normal anterior direction.

Common carotid artery stenosis, near its aortic ostium, is diagnosed as mentioned by Pourcelot, by finding turbulence low in the common carotid artery. This finding may explain a bruit low in the

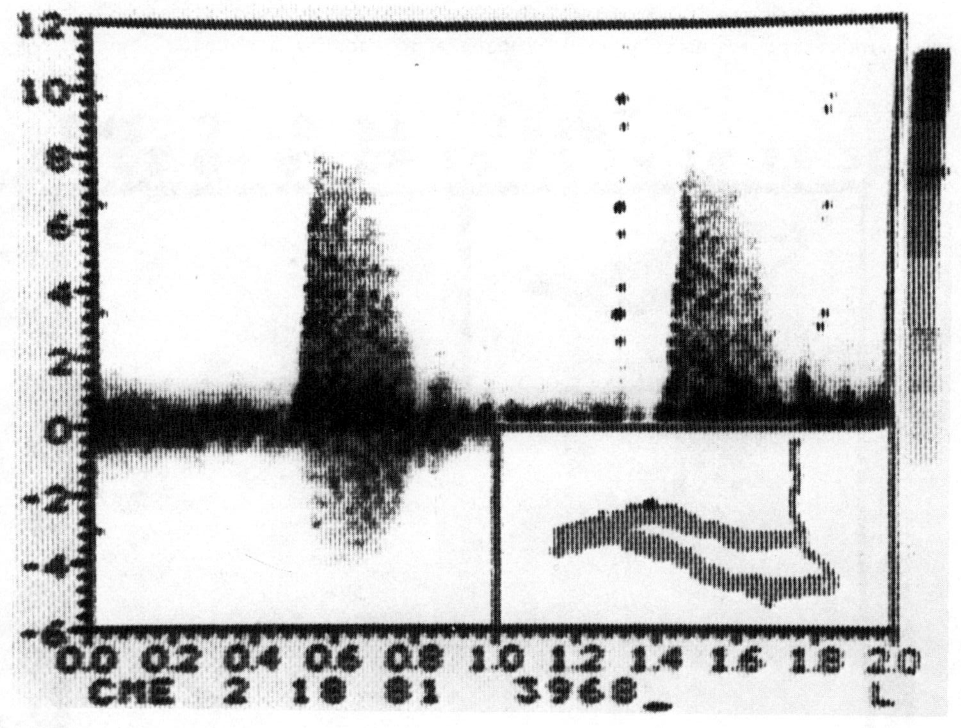

FIG. 7.28A   Full capability demonstration of sufficient information to
make a visual diagnosis of external carotid stenosis high frequency
represented is 8 kHz with spectral broadening and reversing turbulence

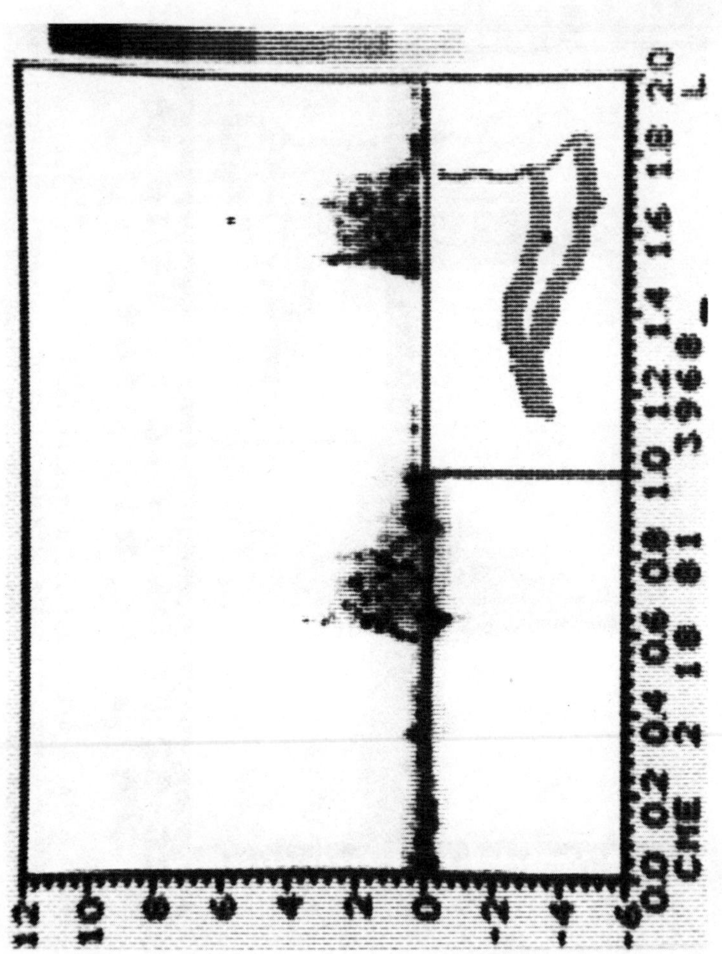

FIG. 7.28B Secondary feature of external carotid stenosis represented on the downstream spectrum by whiskering.

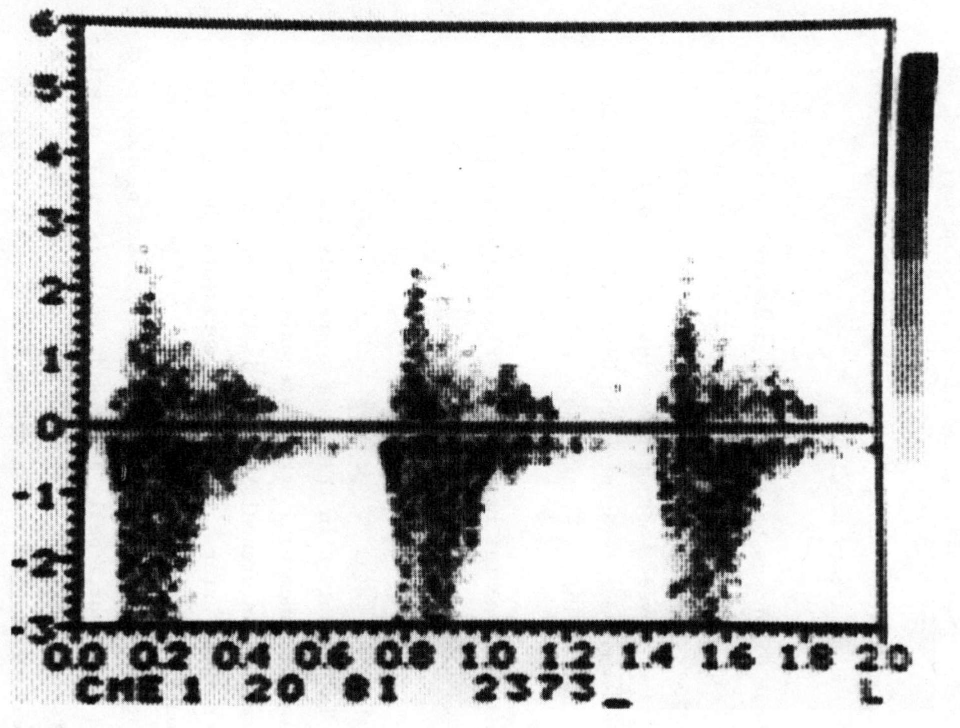

FIG. 7.29   Spectrum taken from the left subclavian artery illustrating diagnostic features of subclavian stenosis. The normal backflow phase is limited
and considerable spectral broadening and reversal of the spectrum indicate
gross turbulence causing a bruit. A slightly high frequency is not fully
displayed. The movement of blood in the normal direction is indicated below
the zero base line because the sound beam is directed against the normal
flow direction.

212

neck. Total occlusion of the common carotid artery produces scattered images of the external and other collaterals and no clear carotid bifurcation can be imaged. Occasionally the bifurcation remains patent and flow is reversed in the internal carotid artery but with low velocities requiring great skill to be identified (Spencer and Reid, 1981).

7.3.2.7   Subclavian And Innominate Artery Obstructions And Vertebral
          To Subclavian Artery Steal

Criteria for these diagnoses have been well presented in chapter 5. Subclavian artery stenosis (Fig. 7.29) and vertebral to subclavian artery steal are frequently found if routine Doppler examination of these arteries is employed. Most of these lesions found in patients appear not to be symptomatic (Spencer and Reid, 1981). The technique of determining their clinical significance was disclosed earlier in this chapter.

To understand why brachial reactive hyperemia produces a steal tendency, the equation:

$$R = \frac{\Delta p}{q}$$

from chapter 3 is helpful. R in this case represents the resistance of the subclavian or innominate artery stenosis. Δp represents the pressure drop across the stenosis when p_1 is the aortic arch pressure and p_2 is the brachial artery pressure. q represents the volumetric blood flow through the stenosis.

Before inflation of the cuff Δp is elevated because p_2 is decreased while q is proportionally decreased. If Δp is great a vertebral flow reversal exists at all times throughout the cardiac cycle because the pressure at the basilar artery level (p_b) is greater than p_2 throughout the heart cycle. If a further drop in p_2 is produced by reactive hyperemia in the arm the steal is increased and Doppler detected vertebral artery velocities increase. If a latent steal situation exists p_b and p_2 are approximately equal differing only in phase and time of arrival of the pulse. If p_2 is further reduced by reactive hyperemia a biphasic vertebral artery signal is converted to a continuous steal (Fig. 7.29). A direction sensitive

Doppler signal is not needed to diagnose vertebral artery to subclavian steal when using the reactive hyperemia test.

REFERENCES

Barnes, R.W. and Wilson, M.R. (1975). Doppler ultrasonic evaluation
of cerebrovascular disease. The University of Iowa, Iowa City,
Iowa.

Brockenbrough, E.C. (1976). Screening for the prevention of stroke:
use of a Doppler flowmeter. Information and education resource
support unit, Washington/Alaska regional medical program.

Brockenbrough, E.C. (1981). The periorbital collateral arteries. In
Cerebrovascular evaluation with Doppler ultrasound (Edited by
M.P. Spencer and J.M. Reid). pp. 143-156, Martinus Nijhoff, The
Hague-Boston-London.

Fish, P.J., Wilson, I.M., Holt, B., and Walters, D. (1978). Multi-
channel pulsed Doppler imaging: Measurement accuracy and beam/
vessel angle estimation. In Ultrasound in medicine (Edited by
D. White and E.A. Lyons), Volume 4. pp. 363-376, Plenum Press, New
York-London.

Itti, R., Pottier, J.M., and Pourcelot, L. (1978). Exploration du
système cardio-vasculaire au moyen des radio-isotopes et des ultra-
sons. Aspects complementaires de methodes et interêt de leur association. These pour le doctorate en medecine, Académie d'Orleans-
Tours, Université François Rabelais, France.

Nippa, J.H., Hokansen, D.E., Lee, D.R., Sumner, D.S., and Strandness,
D.E. (1975). Phase rotation for separating forward and reverse
blood velocity signals. IEEE, Transactions on Sonics and Ultrasonics SU 22, 340-346.

Pourcelot, L. (1976). Diagnostic ultrasound for cerebral vascular
diseases. In Present and future of diagnostic ultrasound (Edited
by I. Donald and S. Levi). pp. 141-147, Kooyker Sci., Rotterdam.

Reid, J.M., Davis, D.L., Ricketts, H.J., and Spencer, M.P. (1974).
A new Doppler flowmeter system and its operation with catheter
mounted transducers. In Cardiovascular applications of ultrasound
(Edited by R.S. Reneman). pp. 183-192, North-Holland/American
Elsevier, Amsterdam-London-New York

Reneman, R.S. and Hoeks, A.P.G. (1977). Continuous wave and pulsed
Doppler flowmeters - a general introduction. In Echocardiology
with Doppler applications and real time imaging (Edited by N. Bom).
pp. 189-205, Martinus Nijhoff, The Hague-Boston-London.

214

Reneman, R.S. and Spencer, M.P. (1979). Local Doppler audio spectra in normal and stenosed carotid arteries in man. Ultrasound Med. Biol. <u>5</u>, 1-11.

Spencer, M.P. and Arts, T. (1981). Carotid stenosis and the Bernoulli principle. Fed. Proc. <u>40</u>, 444.

Spencer, M.P. and Reid, J.M. (1979). Quantification of carotid stenosis with continuous-wave (CW) Doppler ultrasound. Stroke <u>10</u>, 326-330.

Spencer, M.P. and Reid, J.M. (1980). Full capability Doppler diagnosis. In <u>Pathophysiology and pharmacotherapy of cerebrovascular disorders</u> (Edited by E. Betz, J. Grote, D. Heuser, and R. Wullenweber). pp. 59-63, Verlag Gerhard Witzstrock, Baden-Baden, Köln, New York.

Spencer, M.P. and Reid, J.M. (1981). <u>Cerebrovascular evaluation with Doppler ultrasound</u>, Martinus Nijhoff, The Hague-Boston-London.

Spencer, M.P., Reid, J.M., Davis, D.L., and Paulson, P.S. (1974a). Cervical carotid imaging with a continuous-wave Doppler flowmeter. Stroke <u>5</u>, 145-154.

Spencer, M.P., Reid, J.M., and Paulson, P.S. (1974b). Diagnosis of carotid artery disease and cerebral vascular insufficiency with Doppler angiography and ophthalmic artery sonography. In <u>Cardiovascular applications of ultrasound</u> (Edited by R.S. Reneman). pp. 249-265, North-Holland/American Elsevier, Amsterdam-London-New York.

Von Reutern, G.M., Budingen, H.J., and Freund, H.J. (1976a). Dopplersonographische Diagnostik von Stenosen und Verschlussen der Vertebral-Arterien und des Subclavian-steal Syndromes. Arch. Psychiatr. Nervenkr. <u>222</u>, 209-222.

Von Reutern, G.M., Budingen, H.J., Hennerici, M., and Freund, H.J. (1976b). Diagnose und Differenzierung von Stenosen und Verschlussen der Arteria Carotis mit der Doppler-Sonographie. Arch. Psychiatr. Nervenkr. <u>222</u>, 191-207.

White, D.N. and Curry, G.R. (1978). A comparison of 424 carotid bifurcations examined by angiography and the Doppler echoflow. In <u>Ultrasound in medicine</u> (Edited by D.N. White and E.A. Lyons), Volume 4. pp. 363-376, Plenum Press, New York-London.

ACKNOWLEDGEMENTS

The work reported here has been supported by the USA Department of Health and Human Services, Research Grant # R01 HL 19341-05.

CHAPTER 8

CHAPTER 8
Non-invasive Demonstration of Disease at the Carotid Bifurcation by Ultrasound

R. G. Gosling, M. G. Beasley, *and* R. R. Lewis

8.1   INTRODUCTION

Atheromatous involvement of the carotid bifurcation in patients with
ischemic cerebrovascular disease is well recognized. Seventy-five per
cent of lesions in such patients are in the extracranial cerebral
vessels (Fields et al, 1968) and in thirty-four per cent a single
lesion is present at the origin of the internal carotid artery,
which is the site most frequently involved (Wass et al, 1968). Some
lesions are clinically silent while others are associated with transient ischemic attacks, strokes and mental impairment. One important
difference is likely to be the production of platelet emboli, but
this is still to be conclusively demonstrated. One of the difficulties in elucidating these problems is knowledge of the extent and
natural progression of asymptomatic occlusive disease of the carotid
arteries in the general population. To this end many investigators
are now using non-invasive imaging, both directly and also indirectly
by observing the hemodynamic consequences of such disease. It is
anticipated that such methods will provide a firm foundation for
better management of symptomatic and asymptomatic lesions. However,
the experience of different laboratories with the same method often
seems to produce conflicting results.

Evaluation of the different methods for assessing carotid artery
disease is complicated by the absence of techniques which can at
present accurately assess arterial disease other than by surgical

exposure or post mortem examination. The use of contrast arteriography has been developed to a high standard since the first carotid arteriogram by Moniz in 1927 and the resolution of X-rays is such that clear pictures are obtained in the views taken. However, lesions present in different planes to the X-ray projection can be missed and 40% of carotid bifurcations showing ulceration at surgery may not be demonstrated (Gomensoro et al, 1973; Edwards et al, 1979). Thus apart from the ethical question of subjecting volunteer subjects or patients to arteriography because of its morbidity and mortality (Wass et al, 1968), the method cannot be regarded as a wholly accurate basis for comparison. New methods must be sought, preferably non-invasive and atraumatic. Clearly they must be at least as accurate as arteriography and should:

- indicate the position of the carotid bifurcation and show which of the carotid arteries is being assessed
- differentiate between a completely occluded and severely stenosed lumen
- show bilateral disease
- show minor lesions
- show the presence of disease in any arterial segment (e.g. distinguish between lesions in the internal carotid artery in the neck from those at the siphon)
- distinguish between calcium in atheromatous plaque and calcium deposit in the artery wall not associated with atheroma intruding into the lumen.

Further, any new methods should be better than arteriography in:
- showing the presence of ulceration in the arterial wall or occluding plaque
- demonstrating abnormal blood flow in any arterial segment which may be a precursor of arterial disease.

    In the last ten years several new methods of assessment have been proposed, and other than those using ultrasound these have included:
- assessment of frequency, amplitude and position of signals from bruits in the neck (Duncan et al, 1975; Chapman et al, 1980)
- thermography (Wood, 1965)
- thermometry (Shapiro et al, 1970)

- occuloplethysmography - there are two techniques, one for determining the systolic blood pressure in the ophthalmic artery with
instruments based on that developed by Gee (Gee et al, 1976; Gross
et al, 1977) and the other for comparing the amplitudes and arrival times of pressure pulses in the ophthalmic artery of each eye,
as well as comparing the pulse arrival times between the eye and
the external carotid artery at the ear, as developed by Kartchner
and co-investigators (1976).

All of the above methods have inherent physical limitations and
some are measuring secondary and tertiary parameters of the disease.
It is the authors' opinion that ultrasound offers the technique most
likely to succeed, especially since both echo and Doppler investigations are completely non-invasive, atraumatic and safe. The latter
is thus far the only means of detecting abnormal flow patterns by
direct measurement of erythrocyte velocities at different parts of
a vessel lumen either separately or simultaneously, as reported by
Reneman and Spencer (1979). The extent to which this advantage can be
utilised for determining the presence of sub-clinical showers of emboli is now under investigation by several different research groups.
The purpose of this chapter is to present our experience of using
continuous wave Doppler-shift ultrasound to attempt to satisfy the
eight requirements listed above.

8.2   APPARATUS

8.2.1   Spectral Analysis Of Doppler Signals

In our technique a directional Doppler blood-velocity meter, such
as a 4 MHz or 8 MHz Sonicaid BV 380, a 10 MHz Parks 806 or 5 MHz
Parks 705 is used to insonate arteries through the skin. In these
instruments the transmitter and identical receiver crystals are
mounted coplanar at the tip of a pencil shaped holder and receive
ultrasound scattered by any elastic bodies in the lumen, principally
the erythrocytes. This backscattered ultrasound is subjected to
immediate time compression spectral analysis (Coghlan et al, 1974;
Coghlan and Taylor, 1979) and the complete range of Doppler-shifted
frequencies detected is displayed in the form of a sonogram (Fig. 8.1)

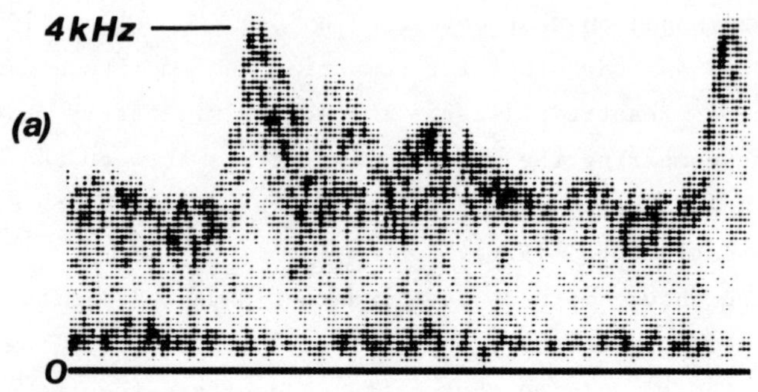

FIG. 8.1   Sonogram display of the Doppler shifted
signals obtained by insonation of the internal
carotid artery. Doppler-shift frequency is on
the y-axis, time on the x-axis and the trace
blackness relates to the signal amplitude.

as the signals are simultaneously monitored with earphones. The sonogram outline or "waveform shape" represents the Doppler-shift of
those erythrocytes moving with the highest velocity at any instant,
that is the streamline of fastest flow. The presence of other blood
corpuscles moving with lesser velocities are shown by 80 display
units whatever the frequency analysis range selected for the display,
which may be 0-2.25 kHz, 0-4.5 kHz or 0-9.0 kHz. These display units
appear as 'dots' printed in a vertical sequence on the screen of the
memory oscilloscope. One vertical sweep is printed every 6.8 milliseconds and the blackness of each 'dot' is related to the number of
corpuscles moving in the particular velocity range for that display
unit, i.e., each unit has a Doppler-shift bandwidth equal to the
frequency range selected divided by 80, so that if the 0-4.5 kHz
range is used for analyzing the Doppler signals received, then each
'dot' represents signals from the number of corpuscles moving with
velocities giving rise to Doppler-shifts in 56.25 Hz intervals.
Sonograms composed in this way, therefore, give information about
the blood velocity distribution at all points across the insonated
lumen, in steps of velocity equivalent to 56 Hz, every 6.8 ms of
each cardiac cycle. This apparatus is known as the Spectrascribe and

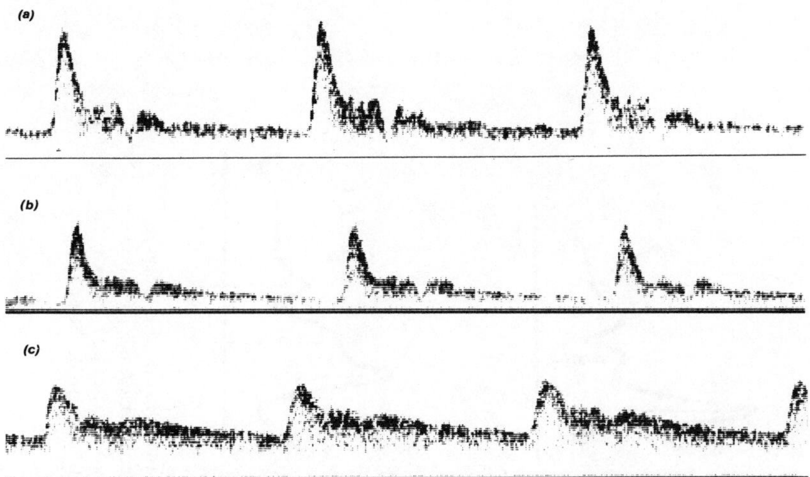

FIG. 8.2   Typical normal sonograms for three successive heart
   beats from the (a) common carotid artery at the base of
   the neck, (b) external and (c) internal carotid arteries
   at the angle of the jaw. The internal can be distinguished
   from the external carotid artery by the higher diastolic
   flow velocity relative to that in peak systole.

the Spectral Analyzer in this system is now commercially available
as a separate module, from Kranzbühler.

As can be seen from Fig. 8.1 the range of blood velocities and
the distribution of erythrocytes within the range varies throughout
the cardiac cycle. It also varies in different arteries as has been
reported by Nimura et al (1974). We have found that the range has a
typical pattern in 'normal' subjects which is characteristically
different for the internal, external and common carotid arteries
(Fig. 8.2) thus enabling the examiner to identify which artery is
being insonated. These 'normal' patterns are changed by age (Fig. 8.3)
and by the presence of atheroma in any part of the carotid pathway,
including the intracranial portion. Disease is therefore detected
by changes in sonograms from the carotid and supraorbital arteries
(Figs. 8.4 and 8.5). In general, the advantages of a sonogram display over other methods of displaying Doppler signals, such as zero-
crossing frequency averaging, are fourfold:

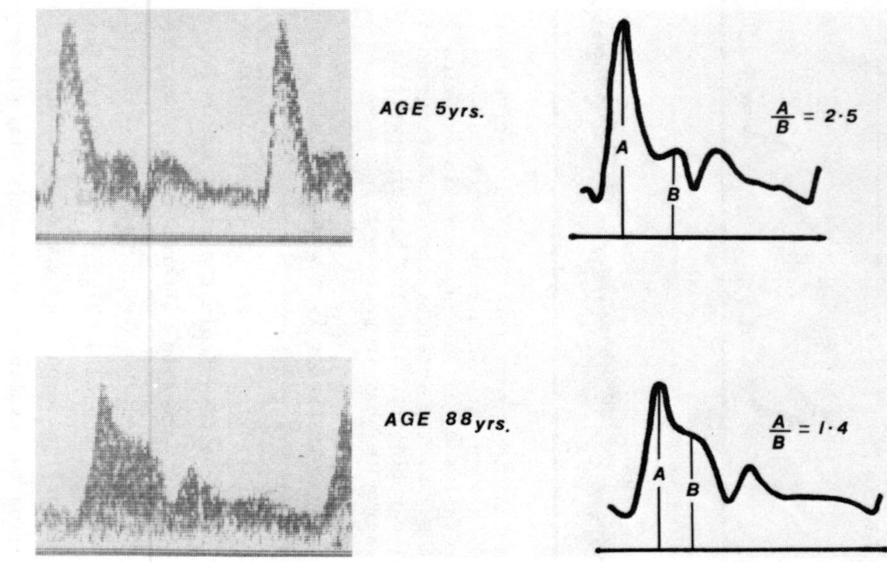

FIG. 8.3   Sonogram display of Doppler signals from the common carotid artery in two normal subjects, and drawing of sonogram waveform shape to illustrate the two principal peaks in systole, A and B. Note the difference of the peak heights' ratio, A/B, with age.

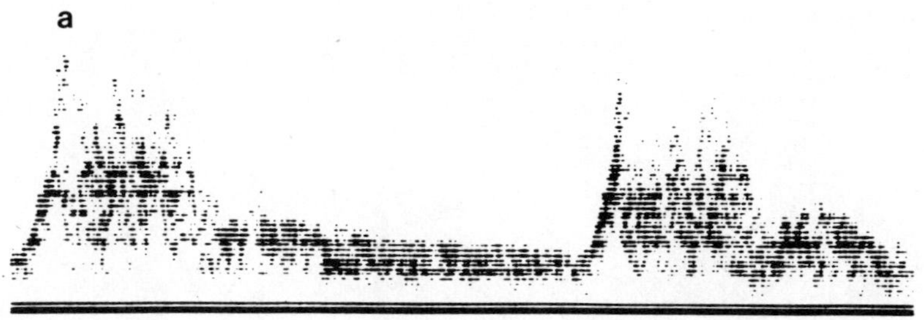

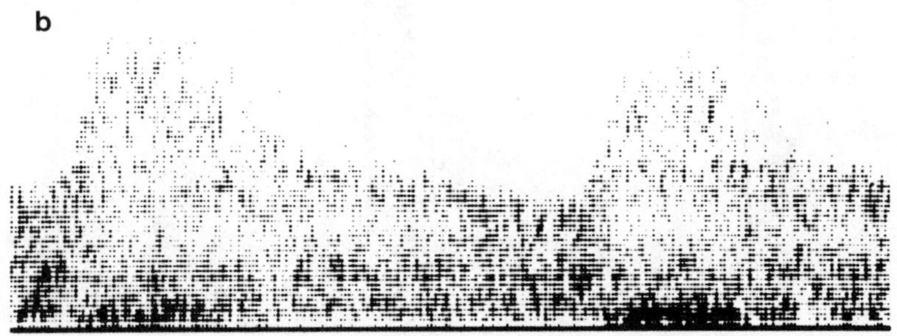

FIG. 8.4 Sonograms illustrating turbulence: the normally
smooth waveform shape is fragmented due to sudden random
increases in Doppler-shift frequencies from the (a) common
carotid and (b) internal carotid arteries.

- the full spectral distribution is displayed and this, which is
  characteristic for most arterial sites in a 'normal' subject
  (Figs. 8.2 and 8.3), alters with disease (Fig. 8.5), including
  turbulence (Fig. 8.4), but still permits distinction between
  common, internal and external carotid arteries.
- a fast response is obtained to changes of erythrocyte velocities
  (Fig. 8.6).
- good discrimination of signals may still be achieved even when
  the signal to noise ratio is low (Fig. 8.7).
- artefacts are shown clearly; for example vessel wall movement,
  multiple flow signals and radio interference (Fig. 8.8).

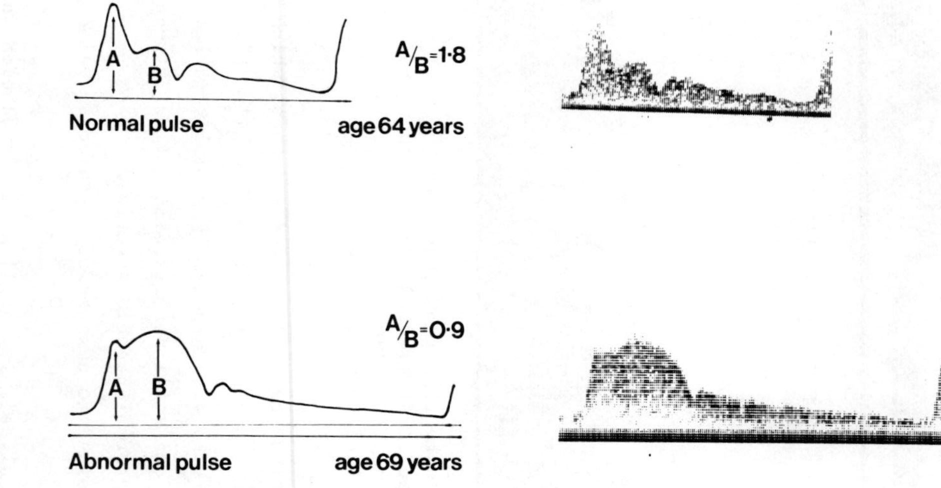

FIG. 8.5 Sonograms from the supraorbital arteries in two men, one aged 64 years with no signs or symptoms of vascular disorder and the other aged 69 years with occlusive disease in the ipsilateral internal carotid artery. The trace of the sonogram outline shows the difference in A/B ratios.

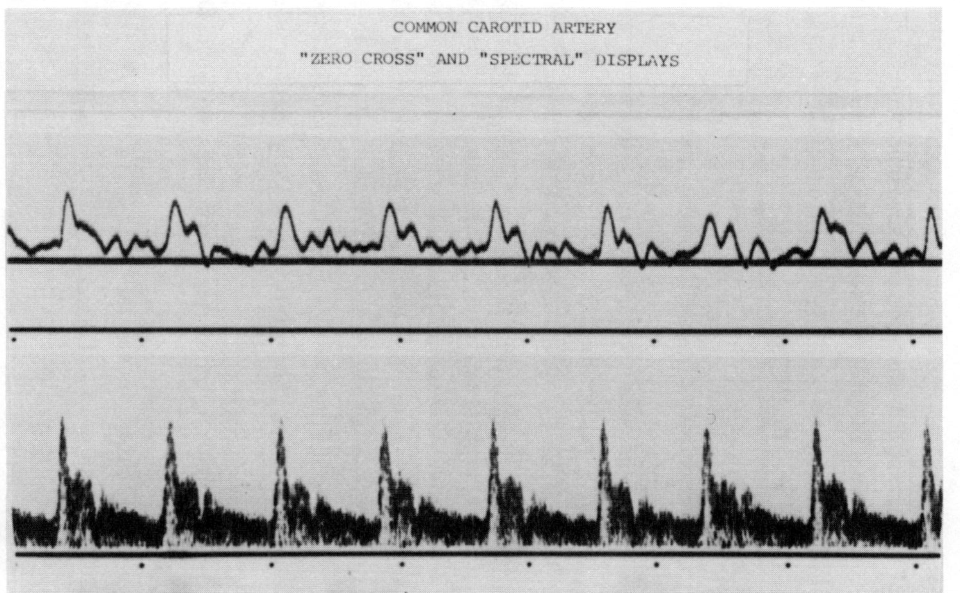

FIG. 8.6   Upper trace: output of voltage controlled oscillator driven by a zero-crossing frequency meter receiving Doppler-shift signals from an insonated common carotid artery, displayed on one channel of the Spectrascribe. Lower trace: simultaneous direct spectral display of the same Doppler signals on the second channel of the Spectrascribe.

224

Previous to the application of sonogram analysis to Doppler signals
obtained from the carotid arteries, the use of continuous wave
ultrasound to detect lesions in these arteries was only reliable for
severe disease, that is occlusions of greater than about 85% in diameter of the vessel lumen (Brockenbrough, 1970; Machleder, 1973).
However, as will be shown below, by measuring that ratio of the two
peaks, A/B, seen in the systolic part of the sonogram waveform
derived from the common carotid and supraorbital arteries, it has
been possible to devise a simple test which can detect many of the
smaller lesions (Baskett et al, 1977).

8.2.2   Operator Interaction

The most noticeable aspect of using real-time and continuous spectral
display of signals from any vessel is the simultaneous information
the operator receives from hand (position of incident beam in the
patient), ear (the Doppler sounds) and eye (the sonogram display).
As a result comparatively little training is necessary to identify
and trace the course of the carotid arteries in the neck because of
the recognizable differences in sonograms from the common, external
and internal carotid arteries as shown in Fig. 8.2. Any Doppler
blood-velocimeter with good directional resolution in conjunction
with a two channel spectral analyzer also enables the patterns of
sonograms from the carotid arteries to be easily demonstrated even
though it is not always possible to avoid insonating the internal
jugular vein (Fig. 8.9). Furthermore, in the presence of disease
the vessel geometry is often distorted so that it is possible for
the external and internal carotid arteries to overlie and twist
around each other immediately distal to the carotid bifurcation.
The presence of such distortion can be detected by inspection of
the sonogram display as shown in Fig. 8.10, where the external carotid artery sonogram pattern can be seen within that from the internal
carotid artery. This would not be discernible by any attempt at processing the Doppler signals before display, as is done with a zero-
crossing frequency meter (Lunt, 1975), a maximum frequency follower
(Sainz et al, 1976), or a mean frequency follower (Smallwood and
Brown, 1978).

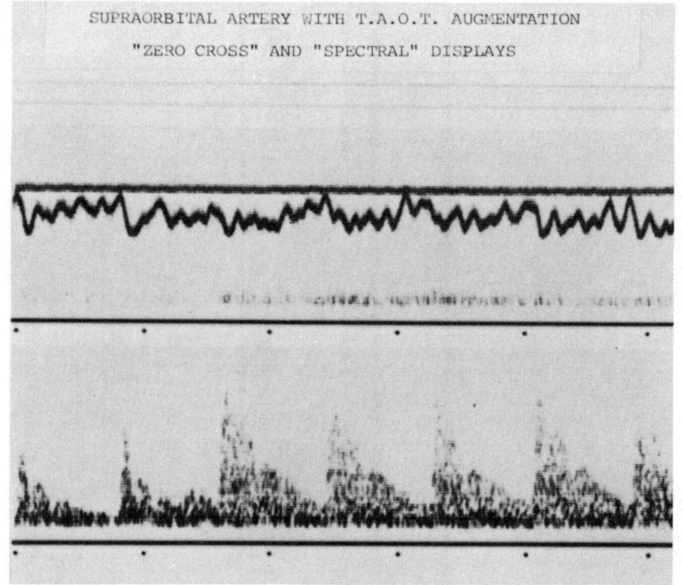

FIG. 8.7   Doppler-shift signals from an insonated supraorbital artery obtained during compression of the ipsilateral temporal artery. Compare display in the upper
trace, taken from the zero-crossing frequency meter, with
that of the lower trace showing direct spectral analysis.
Compression was applied after the second heart beat.

8.2.3 <u>Direct Imaging Of The Carotid Bifurcation</u>

A continuous wave ultrasound vessel imaging system has been designed
and constructed in our group, essentially as an "add-on" unit to
the Spectrascribe. This accepts from the Spectrascribe time compression analyzer all the information from the Doppler frequency/ampli-
tude 'display units', which is in digital form, and uses this in a
microprocessor system (6800) to determine the bright up criterion
for the image formation (Coghlan and Taylor, 1978a). Thus, simply by
writing the appropriate software the use of different bright up criteria may be explored. We are currently comparing three programs
based on the power within the frequency spectrum in an effort to

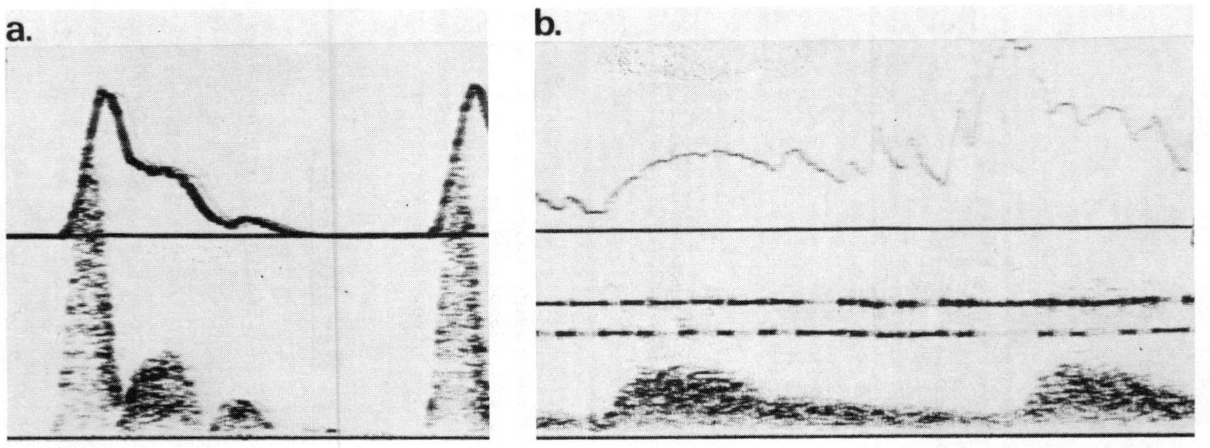

FIG. 8.8   Comparison of Doppler signals processed simultaneously by zero-crossing frequency averager (top traces) and spectral analyzer (lower traces)

(a) Measurement of peak velocities at various phases of the cardiac cycle is more difficult with zero-crossing frequency meter because of the smoothing due to time constants inherent in the circuitry (here $\tau = 0.045$ sec).

(b) Sonogram waveform is unaffected by radio frequency interference.

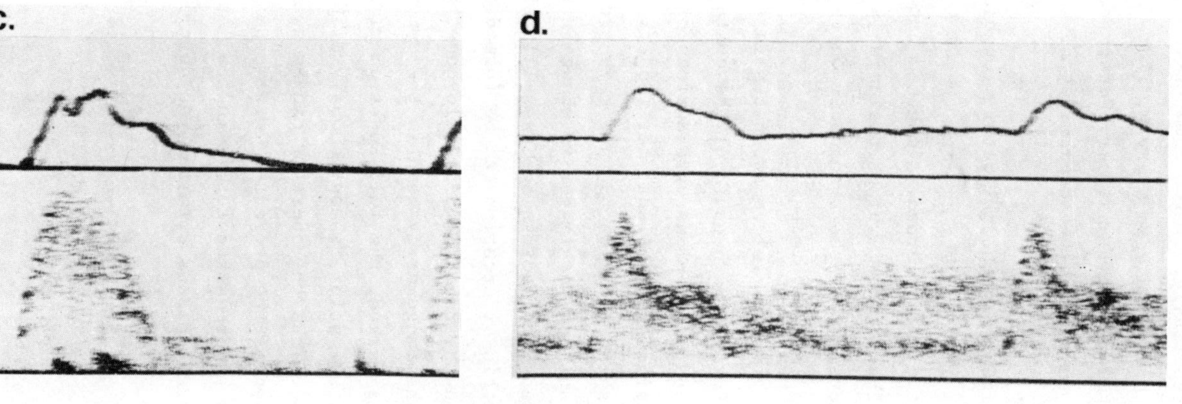

FIG. 8.8 (continued)

(c) Wall velocity and probe movement can both introduce strong low frequency signals which alter the zero-cross waveform shape, but leave the maximum frequency outline of the sonogram unchanged.

(d) Venous and arterial flow can be separately discerned on the sonogram despite being in the same direction, whereas the zero-cross processing yields a single apparently damped arterial signal.

228

obtain better definition of the edge of the lumen (Coghlan, 1979)
as the transducer is moving across the neck. These include a cross-
correlation analysis which essentially permits comparison of the
distribution of amplitude with frequency in every 4, 8 or 32 of the
6.8 ms electronic sweeps of the spectral analyzer. Future research
will include a comparison of the diagnostic capability of velocity
as opposed to power criteria.

The imager has been described in technical detail (Coghlan and
Taylor, 1978a; Coghlan, 1979) and consists essentially of 4 modules:
- a transducer with two 5 MHz crystals broadly focussed between two
  and four centimeters from their faces, which considerably diminishes interference from movement of the transducer on the skin,
  and a fully directional Doppler velocimeter.
- a scanning arm assembly to give spatial co-ordinates in the Z-X
  plane. The Y plane is not registered because of the contours of
  the neck, so that this method gives planar projections of vessels.
- a two channel time compression spectral analyzer (Spectrascribe).
- a microprocessor, image display oscilloscope and keyboard.

The dual channel display unit of the imager is shown in Fig. 8.11;
by repeated passes of the transducer across the vessel lumen a two
dimensional image of the carotid bifurcation can be formed as shown.
Each image point is marked on the right hand oscilloscope, which is
continually refreshed from the image array held in the computer
memory. As the image point is marked, the full spectrum of blood
velocities corresponding to that point is continuously displayed
on the left hand oscilloscope. Thus the operator can be aware of
abnormal blood velocities even if the projected image appears normal.
In such an event the patient can be asked to lie in the lateral
position and images in different planes obtained, as described
below.

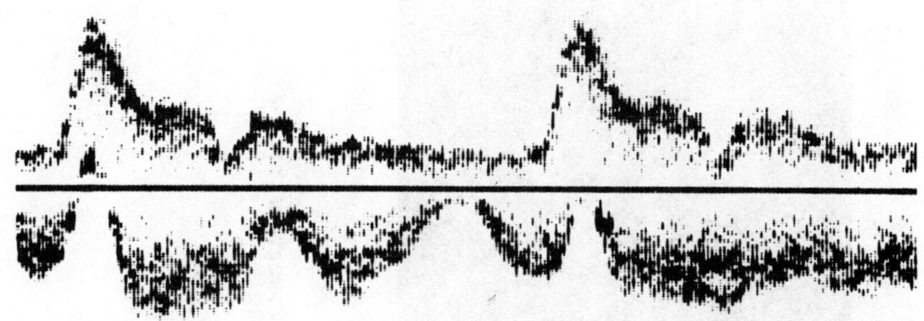

FIG. 8.9   Directional spectral display of Doppler-shift signals
produced by simultaneous insonation of common carotid artery
(upper trace) and jugular vein (lower trace). Venous velocities in the same direction as the arterial flow can be seen
at peak systole in the first heart beat.

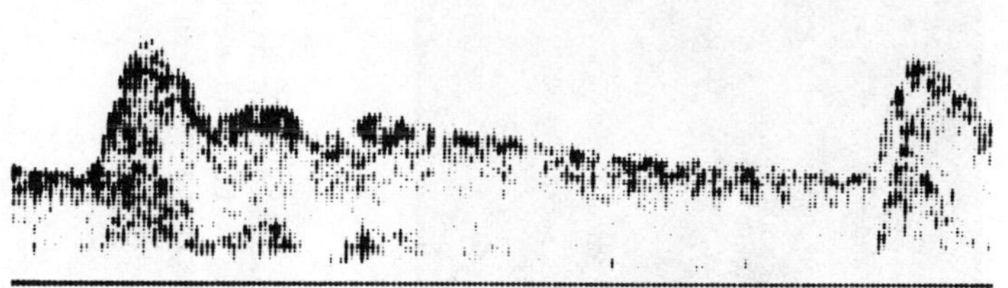

FIG. 8.10   External carotid artery sonogram within that from the
internal carotid artery, seen when one artery overlays the
other so that they are both within the ultrasound beam.

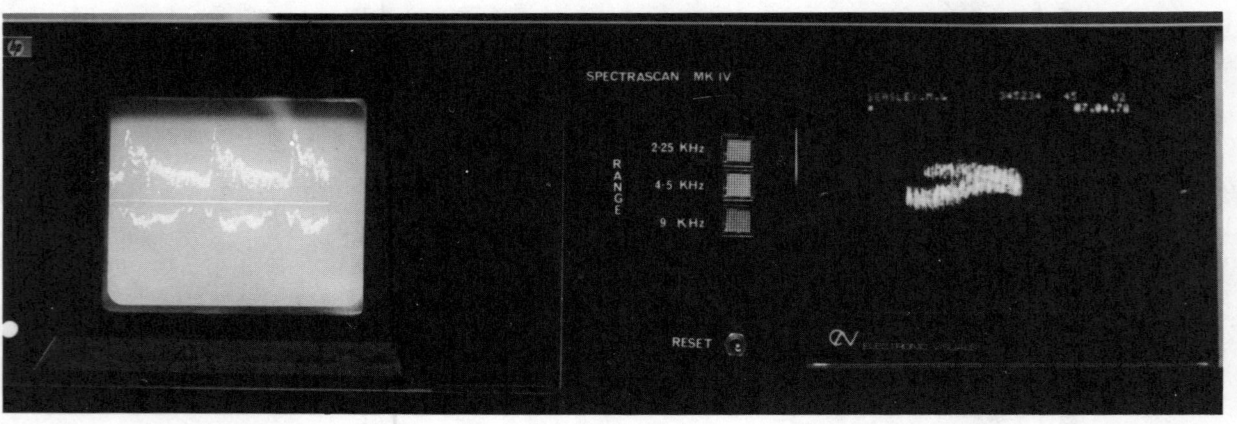

FIG. 8.11   Dual oscilloscope display unit of vessel imager. Sonograms are displayed on the left hand screen simultaneously with each point of the vessel image as it is built up on the right hand screen.

8.3   METHOD

The method is based on a two stage approach, a simple screening test which if positive is followed by imaging of the carotid bifurcation.

8.3.1 <u>Screening Test</u>

All measurements are obtained with the patient lying rested and unsedated on a couch. Sonograms are recorded from the common carotid and the ipsilateral supraorbital arteries (Baskett et al, 1977), using an 8 MHz or 10 MHz blood velocimeter which can separate flow in opposite channels so that venous signals are separated from arterial signals (Coghlan and Taylor, 1978b). It is helpful to monitor the signals with earphones as well as following their display as sonograms. To minimize the effect of respiration and any small changes of the resting pulse rate on the A/B ratio, 20 consecutive pulses are usually analyzed and the average A/B value calculated.

When a sequence of well defined sonograms has been obtained from the supraorbital artery, Brockenbrough's temporal artery occlusion test (TAOT) is applied as illustrated in Fig. 8.12. A positive response to the TAOT or an A/B ratio of less than a certain threshold value, as described in the section on Results, are taken as an indication to proceed to the second stage, that of imaging the bifurcation. This screening test takes 5-10 minutes for each side.

On occasions the TAOT response is neither normal nor positive as illustrated in Fig. 8.12, but 'negative', in that there is no discernible change in the sonogram. This is usually due to the presence of collateral vessels, which can also result in a contradiction between the TAOT response and the A/B ratio value indicative of disease. In either of these cases we have found it is useful to insonate the supratrochlear artery and test the response with compression of the ipsilateral superficial temporal and facial arteries both separately and simultaneously. If doubt still exists the examination sequence of the supraorbital and supratrochlear arteries can be repeated but with compression of the contralateral superficial temporal and facial arteries.

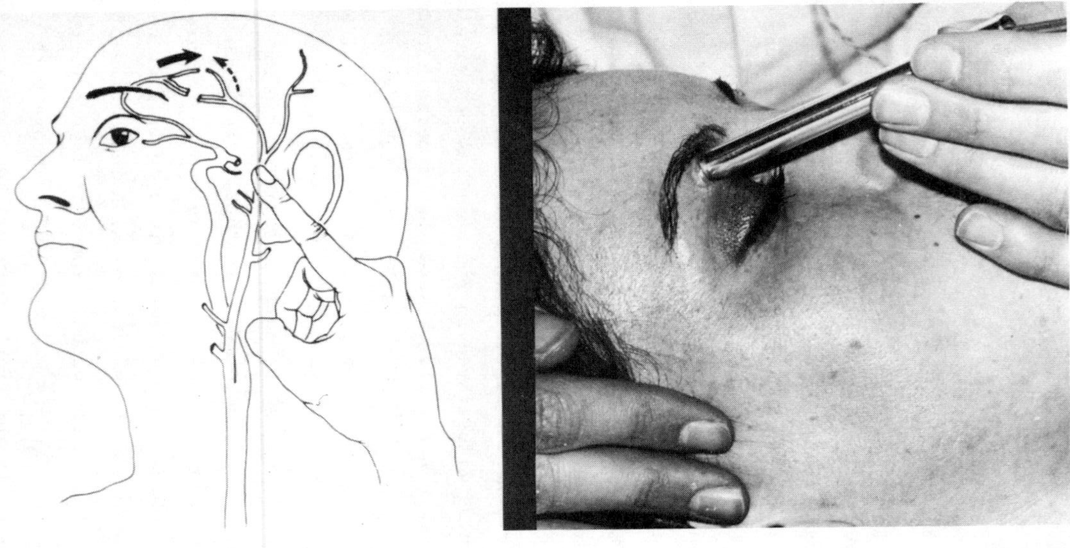

FIG. 8.12   The temporal artery occlusion test (TAOT)
Top right: transducer over the supraorbital artery. Top left: diagrammatic illustration of the anastomic vascular bed in the forehead between branches of the
internal and external carotid arteries. The TAOT is performed by insonating the
supraorbital artery and compressing the ipsilateral superficial temporal artery
against the cranium.

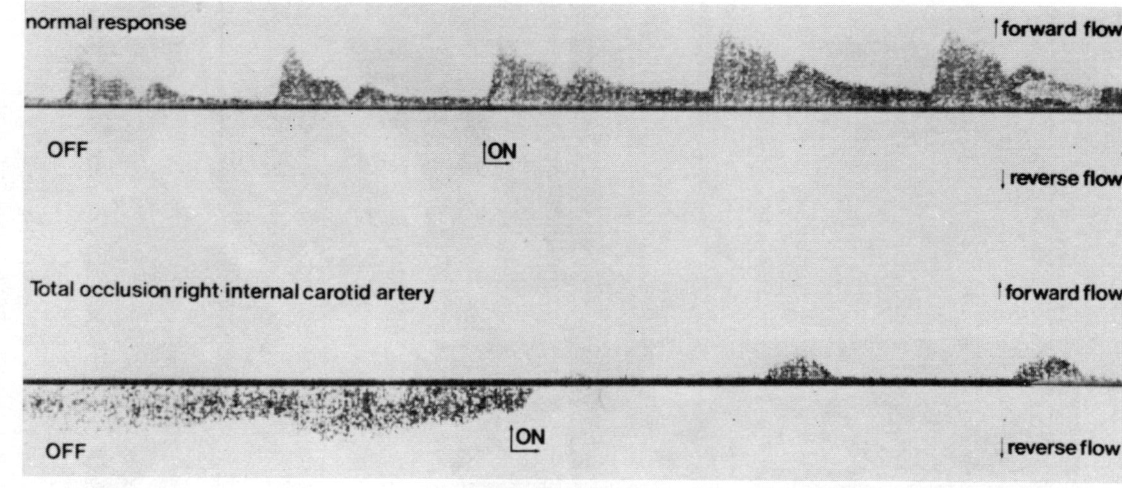

FIG. 8.12 (continued)

Sonograms demonstrating a normal TAOT (top line), with augmentation of flow-velocities during compression. In an abnormal response (bottom line), the blood flow, initially into the orbit, is stopped by compression. In the case illustrated some blood then flows out of the orbit, indicating additional collateral supply to the internal carotid pathway.

We have found from investigating over 200 patients that in most
cases the examiner knows, simply from inspection of the A/B ratio
of the sonograms on the display screen and the TAOT response,
whether or not to use the imager.

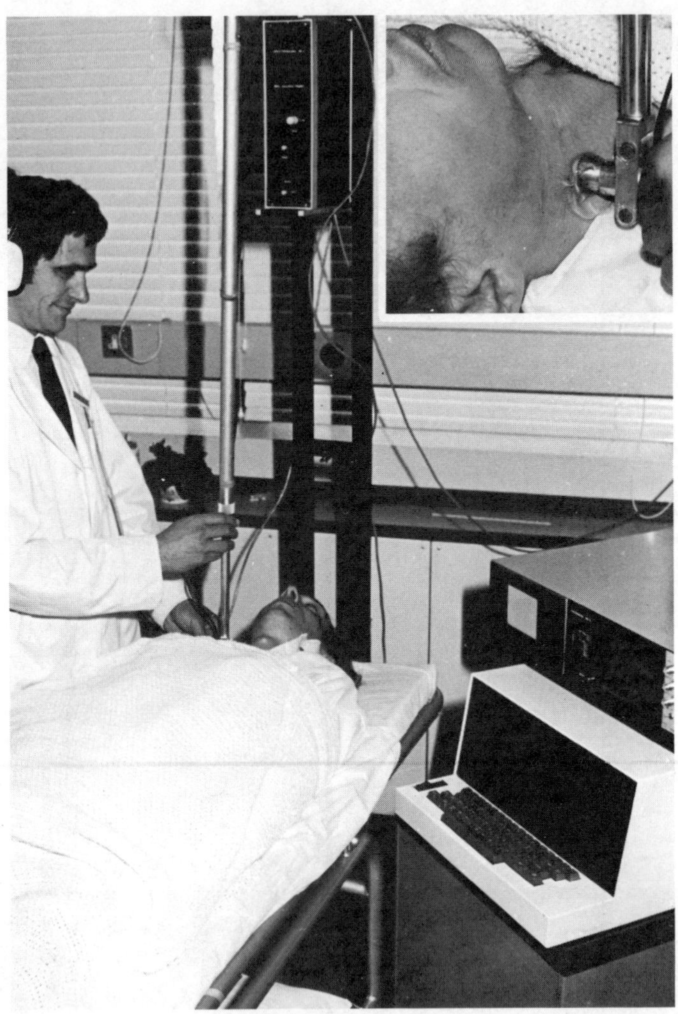

FIG. 8.13   The continuous wave ultrasound imaging
system. The insert shows the transducer and
scanning arm configuration to obtain a lateral
view of the carotid bifurcation.

8.3.2   Imaging The Bifurcation

To produce an image of a lateral projection of the carotid bifurcation, the patient lies supine and the transducer is placed on the
side of the neck so that a beam of ultrasound passes across the
lumen of the underlying artery (Fig. 8.13). For correct registration
of the position of the vessel wall, the direction of the incident
beam must be independent of the neck contour. Therefore, the transducer axis is set normal to the Z-X plane of the scanning arm. An
adequate Doppler-shift in frequency is then obtained by rotating the
plane of the arm so that the beam is set pointing towards the head
at an angle of approximately 60° to the side of the neck. Once the
transducer is fixed in position there is no rotational movement. A
coupling gel is applied to the neck and the transducer is moved vertically across the vessel lumen, thereby crossing the artery in an
antero-posterior direction, until no further signal is heard
in the earphones or seen on the oscilloscope of the spectral analyzer.
These vertical movements are repeated along the length of the common
carotid artery and then the external and internal carotid arteries,
so that a lateral projection of the carotid bifurcation is formed,
as shown in the top trace of Fig. 8.14. An antero-posterior projection may be obtained by examining the patient lying in the lateral
position, when the internal carotid artery appears uppermost in the
projection (Fig. 8.14; lower image). The use of such a procedure has
been described elsewhere (Lewis et al, 1978), and is illustrated in
the Figs. 8.15 and 8.16. This system therefore provides the two
views normally obtained by contrast arteriography, while oblique
views may also be obtained by altering the position of the patient
relative to the transducer. An image can be produced in 10-15 minutes.

8.3.3   Measurement Of Percentage Reduction In Lumen Diameter

To compare the results of the screening tests with arteriography, the
degree of occlusive narrowing shown on the X-ray and direct ultrasound images was determined by measuring the smallest diameter of
the vessel lumen at the stenosis and subtracting this value from

the diameter of the normal internal carotid artery distal to any
post-stenotic or bifurcation dilatation. This reduction was then
expressed as a fraction of the distal diameter to give the percentage of stenosis.

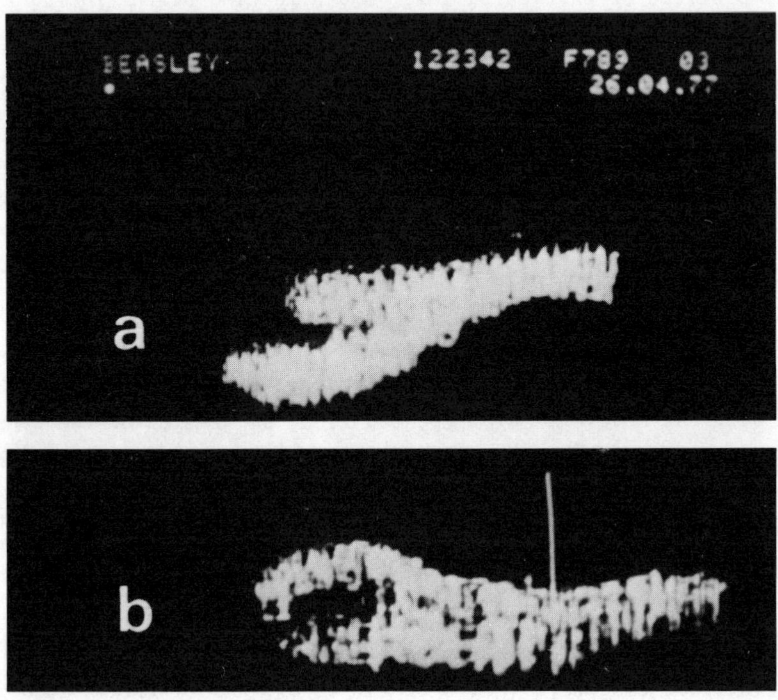

FIG. 8.14   Ultrasound images of the carotid bifurcation
in a healthy young volunteer. (a) lateral view showing
the internal carotid artery inferior to the external
carotid artery. (b) antero-posterior view in which the
internal carotid artery is superior to the external
carotid artery.
Patient details are shown at the top of the screen.

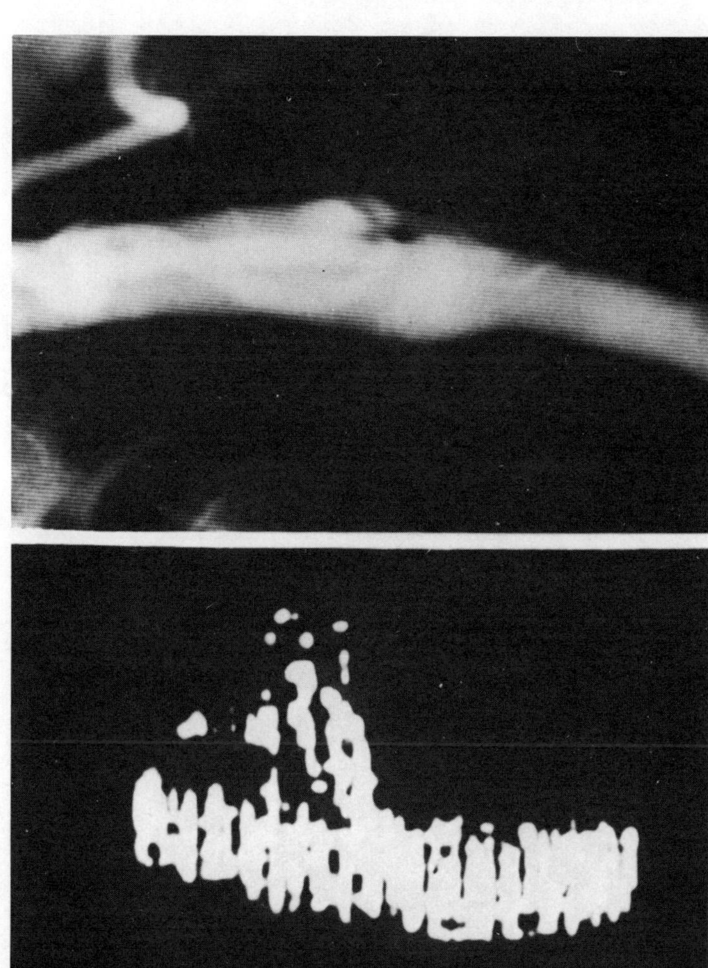

FIG. 8.15 Lateral views of carotid bifurcation by ultrasound and X-ray, from a patient with a bruit in the neck. No definite lesion can be seen in these views.

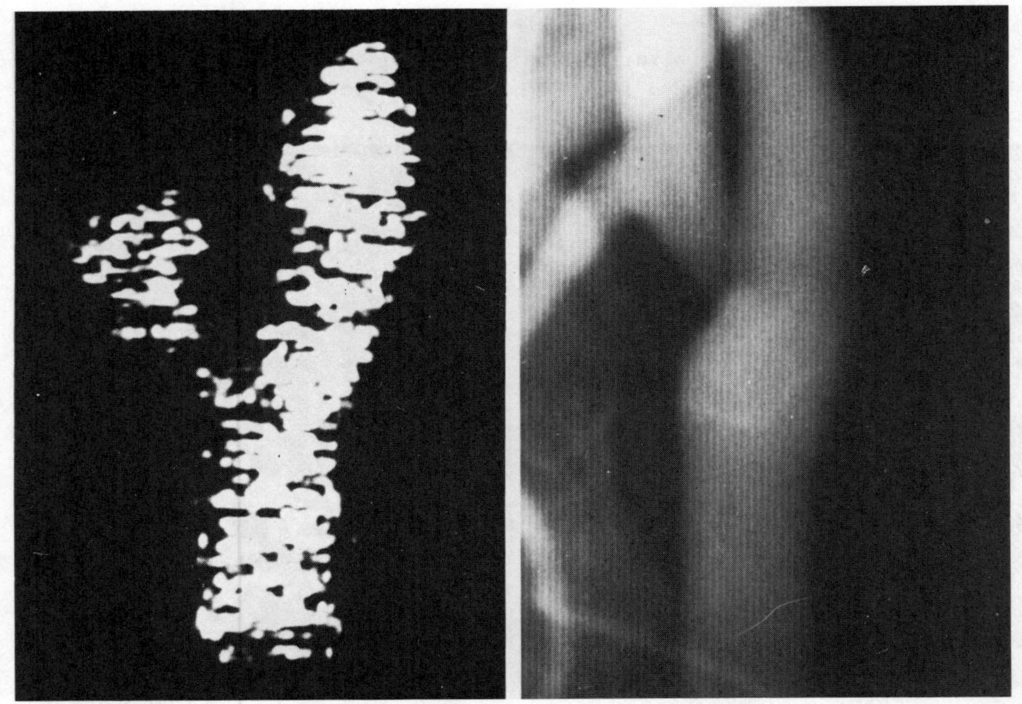

FIG. 8.16   Antero-posterior views of the carotid bifurcation by ultrasound and X-ray, from the patient shown in Fig. 8.15. The external carotid artery now appears to the left of the bifurcation and clearly displays an occlusive lesion.

8.4   RESULTS

8.4.1   <u>Stage One - The Screening Test</u>

The A/B ratio of a sonogram taken from vessels discussed above was
defined in Fig. 8.3. The diagnostic threshold for carotid disease was
first taken from an empirical correlation, shown in Fig. 8.17, between the condition of 103 carotid bifurcations as displayed by angiography and the A/B values from both supraorbital artery and common
carotid artery sonograms. The figure shows that if either the supraorbital artery or the common carotid artery sonogram, or both, show
an A/B value of less than 1.05 disease is likely to be present at
the carotid bifurcation.

Although angiograms are not an ideal standard, for the reasons
expressed in the introduction, at present it is the investigation
with which other techniques are compared. Therefore, following the
suggestion of O'Donnel et al (1980), Table 8.1A sets out such a
comparison in terms of sensitivity and specificity as given by the
decision matrix format defined in Fig. 8.18 (McNeil et al, 1975).
We have now extended this approach by including the TAOT response.
Thus Table 8.1B shows a comparison between the ultrasound screening
test and X-rays using the criterion that carotid bifurcation disease
is present if the TAOT response is positive or either of the A/B
values of supraorbital artery or common carotid artery sonograms is
less than 1.05. It can be seen that the figures for sensitivity and
specificity are similar to those of Table 8.1A and that the number
of bifurcations examined has been doubled, suggesting that the
sample size is sufficient. Table 8.2 shows the analysis of these
figures in terms of the percentage lumen diameter reduction shown by
the angiogram, using six sub-divisions ranging from minimal atheroma,
classified as 0% encroachment on the lumen, to complete occlusion
(100%). As Table 8.2 shows, one case of complete occlusion was mis-
diagnosed. The patient was a 44 year old female who had a right hemi-
paresis and total occlusion of the left internal carotid artery. The
A/B ratios of the left supraorbital artery and common carotid artery
sonograms were normal. However, low velocity signals were obtained
from the left supraorbital artery and the end diastolic velocity in
the left common carotid artery sonogram was lower than that from the

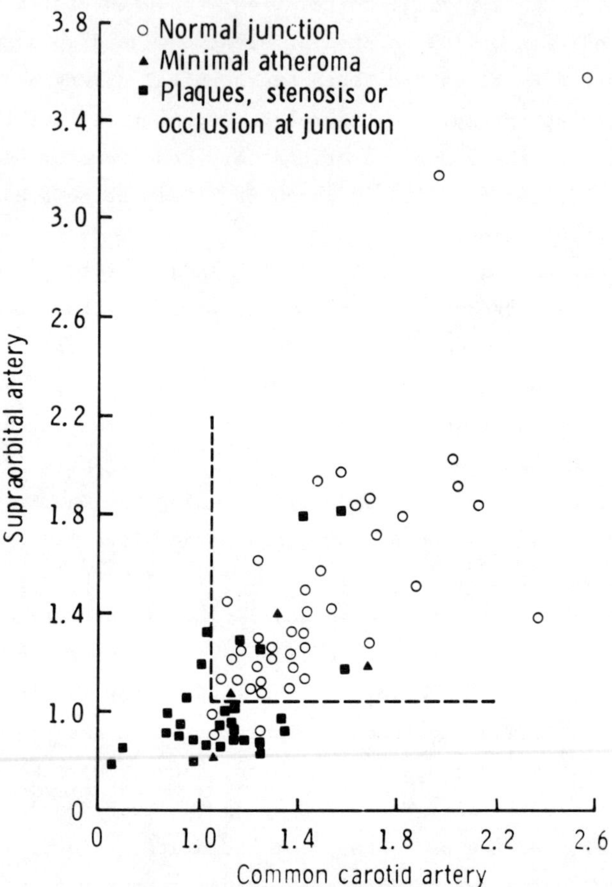

FIG. 8.17   Comparison of angiography results
with A/B ratios from supraorbital and common
carotid sonograms from 103 carotid bifurcations. The broken line is at 1.05 on both
axes.

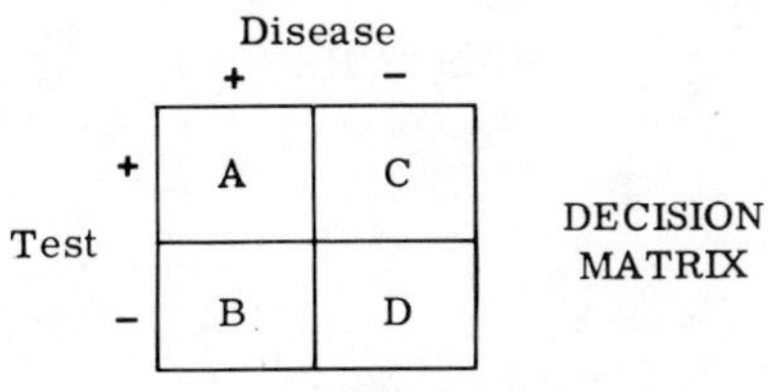

1. SENSITIVITY (True +) = Fraction of patients WITH disease detected by test.

              TP $= A/(A+B)$

2. SPECIFICITY (True −) = Fraction of patients WITHOUT disease detected by test.

              TN $= D/(C+D)$

3 FALSE POSITIVE RATIO $= C/(C+D) = I-TN$

4. FALSE NEGATIVE RATIO $= B/(A+B) = I-TP$

5. POSITIVE ACCURACY $= A/(A+C)$

6. NEGATIVE ACCURACY $= D/(B+D)$

FIG. 8.18   Definition of terms used in the decision matrix

right side. We have subsequently found that such low velocity signals, compared with the contralateral side, are a useful indicator of severe internal carotid artery disease. Apart from this one case, all other 22 cases where the stenosis was greater than 75% were correctly diagnosed from the hemodynamic information of the ultrasonic test. The table also indicates the variable hemodynamic response resulting from lesions of less than 25% stenosis.

Table 8.3A shows that lowering the threshold of the A/B ratio below 1.05 does not help to distinguish the differences in severity of stenotic disease. However, the use of a positive TAOT is discriminatory, and if used together with an A/B value of 0.9 allows detection of disease greater than a 75% stenosis with a true positive ratio (sensitivity) of 96%, as shown in Table 8.3B. Yet, because of the false positives that would occur if this test were used prospectively a positive finding would only indicate a 69% chance of the disease being greater than a stenosis of 75% (i.e. the positive

TABLE 8.1A : Ability of the A/B ratio to detect carotid bifurcation
disease demonstrated by angiography

ANGIOGRAM RESULT

(Excluding Minimal Atheroma)   (Including Minimal Atheroma)

NUMBER OF
BIFURCATIONS: 96   103

		+	−
ULTRA-SOUND 1) TEST	+	34	5
	−	7	50

		+	−
ULTRA-SOUND TEST	+	36	5
	−	12	50

	Excluding	Including
SENSITIVITY (True positive)	83%	75%
SPECIFICITY (True negative)	91%	91%
FALSE POSITIVE	9%	9%
FALSE NEGATIVE	17%	25%
POSITIVE ACCURACY	87%	88%
NEGATIVE ACCURACY	88%	81%

1) Ultrasound test taken as positive if either supraorbital artery
or common carotid artery sonogram A/B value is less than 1.05.

TABLE 8.1B : Ability of the A/B ratio together with the TAOT to
detect carotid bifurcation disease demonstrated by
angiography

ANGIOGRAM RESULT

(Excluding Minimal Atheroma) | (Including Minimal Atheroma)

NUMBER OF BIFURCATIONS: 175

200

	ANGIOGRAM RESULT (Excluding Minimal Atheroma)	
ULTRA-SOUND 1) TEST	+	−
+	53	8
−	11	103

	ANGIOGRAM RESULT (Including Minimal Atheroma)	
ULTRA-SOUND TEST	+	−
+	64	8
−	25	103

	Excluding Minimal Atheroma	Including Minimal Atheroma
SENSITIVITY (True positive)	83%	72%
SPECIFICITY (True negative)	93%	93%
FALSE POSITIVE	7%	7%
FALSE NEGATIVE	17%	28%
POSITIVE ACCURACY	87%	89%
NEGATIVE ACCURACY	90%	80%

1) Ultrasound test taken as positive if either supraorbital artery
or common carotid artery sonogram A/B value is less than 1.05
and/or the TAOT is positive.

TABLE 8.2 : Ultrasound screening test compared to angiography: for degree of stenosis in cervical course of the internal carotid artery. Ultrasound threshold: A/B<1.05 and/or positive TAOT

GROUP No.	PERCENTAGE REDUCTION IN LUMEN DIAMETER	NUMBER IN EACH GROUP	ULTRASOUND TRUE POSITIVE	ULTRASOUND FALSE NEGATIVE	SENSITIVITY (%)
1	0 (minimal atheroma)	25	11	14	14
11	1 - 24	16	9	7	56
111	25 - 49	14	12	2	86
IV	50 - 74	11	10	1	91
V	75 - 99	5	5	0	100
VI	100 (blocked)	18	17	1	94

			ULTRASOUND TRUE NEG.	ULTRASOUND FALSE POS.	SPECIFICITY (%)
NORMALS			111	103	8	93

TABLE 8.3 : Ability of ultrasound screening test to distinguish
severity of disease as defined by angiography

A. For test taken as positive if either supraorbital artery or common
carotid artery sonogram A/B value is less than 0.9.

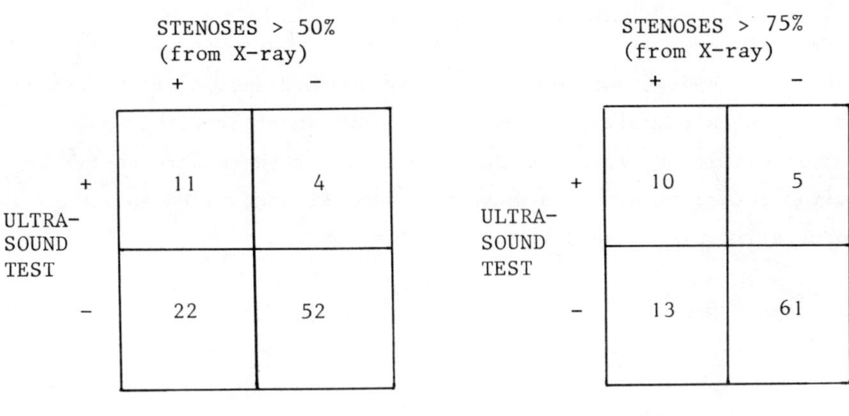

B. For test taken as positive if TAOT is positive and/or A/B value of
either supraorbital artery or common carotid artery sonogram is
less than 0.9.

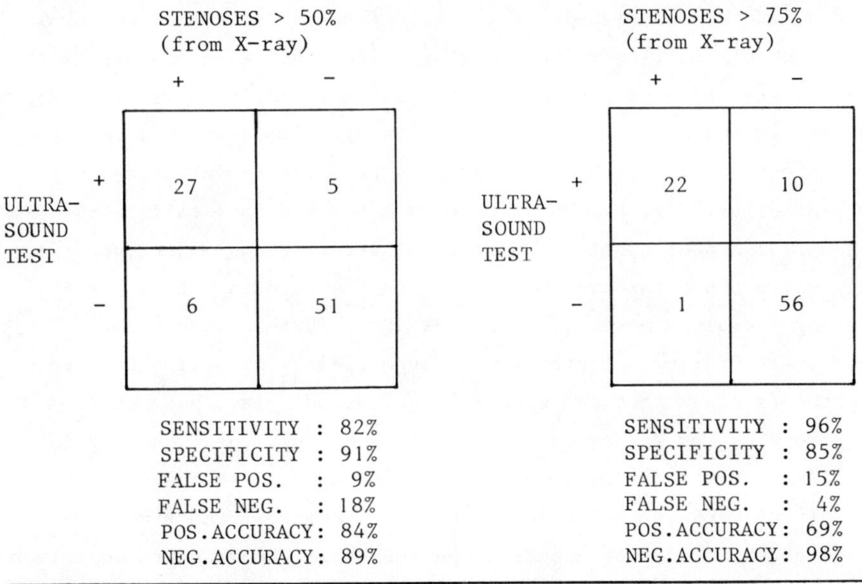

accuracy is only 69%) but an 84% chance of being greater than a 50% stenosis. Thus one may conclude that a sub-division on the basis shown may be reasonably used prospectively to distinguish moderate from major stenosis at the level of 50% diameter occlusion.

8.4.2 Stage Two – Imaging

In Table 8.4 the percentage reduction in lumen diameter of 20 ultrasound images is compared with that from subsequent arteriography performed within 48 hours. In this small series there were no false positives regarding severe stenosis or complete occlusion and one false negative with regard to a small lesion.

8.5 DISCUSSION

8.5.1 Spectral Analysis

An essential feature of the method outlined above is that spectral analysis of the Doppler signals enables the operator to identify artefact and make valid judgements on poor quality signals, that is low signal to noise ratio, often encountered in diseased states of the head and neck arteries. Fig. 8.9 shows not only the separation of arterial and venous signals, which is a function of the directional resolution of the flow-velocimeter circuitry, but also the ease with which reverse flow in a vein can be distinguished from arterial velocities in the spectrally analyzed display. Similarly, and perhaps more importantly, Fig. 8.10 illustrates the discernment of two overlying arteries. Fig. 8.7 shows that an A/B ratio can still be assessed from a poor quality supraorbital artery signal when spectrally analyzed, but not resolved by processing of the data with a zero-crossing frequency meter (Sonicaid BV 380). Further, the normal enhancement of flow velocities with superficial temporal artery compression is clear from the spectral analyzer display, but not from the zero-crossing frequency meter. Fig. 8.6 shows that when a good signal/noise ratio is obtained, resolution of the A/B peaks is possible with the zero-crossing technique. However, the time constants of the processor appear to be such that the standard deviation

TABLE 8.4 : Comparison of lumen diameter reductions on ultrasound and X-ray images from 20 patients

	X-ray image	Ultrasound image
	25%	26%
	29%	43%
	14%	28%
Minor lesions	30%	38%
	28%	42%
	20%	0%
	20%	30%
	90%	95%
	69%	52%
Severe lesions	50%	90%
	79%	78%
Complete occlusion (9 patients)	100%	100%

of the A/B values is high. For example the following values were obtained from the sequence of pulses shown:

Zero-cross: Mean A/B = 1.61 $\pm$ 0.26 (16%)

Spectral Analysis: Mean A/B = 1.71 $\pm$ 0.04 (2%)

Since we were able to attribute significance to a difference in average A/B ratios from 1.05 to 0.9, the 16% error in the zero-cross value would significantly decrease the diagnostic ability of the screening test.

8.5.2   Screening Test

It should be noted that the simple screening test cannot detect the presence of ulceration or of all minor disease, especially minimal atheroma, which therefore appears in Table 8.1 as an increase in the false negatives. This constitutes a limitation of the A/B ratio as a diagnostic parameter. However, as an initial test it is a useful precursor to carotid arteriography since it has such a high specificity, in that the false positive ratio is only 7%. Also, as can be seen from Table 8.2, examination of the angiograms showed that 21 of the 25 bifurcations shown falsely by the test to be negative (normal) had in fact only minor stenoses reducing the diameter of the internal carotid artery by less than 25%.

Table 8.3 shows that the test should be capable of being used prospectively to effectively separate cases of percentage stenosis greater or less than 50%, with a positive and negative accuracy of 84 and 89% respectively. A more fine diagnostic resolution than this would not appear to be possible with this method.

From our results an A/B ratio of 0.9 or less may be expected to produce no false positives, whilst a value of more than 1.59 would always be associated with a normal carotid bifurcation. We believe that an A/B threshold of 1.05 provides the best diagnostic compromise between these two extreme values. Thus we would conclude:
- no hemodynamically significant disease if A/B > 1.05 and TAOT normal
- moderate disease (<50% stenosis) if 0.9 < A/B < 1.05 and TAOT normal
- severe disease (>50% stenosis) if A/B < 0.9 and/or TAOT positive.

The advantage of presenting the results in terms of a decision matrix emphasizes the need to make the choice of threshold value – defining positive and negative results – in relation to the clinical purpose of the test. Thus an A/B ratio threshold of 1.05 and positive TAOT, even though it gives rise to 28% false negatives may still be considered a useful prior screen to arteriography since it should ensure that of 100 subsequent arteriograms, only 7 are likely to show no angiographic evidence of disease (Table 8.1). Further, if the test is to be used for severe carotid artery disease, for example, in patients prior to operations requiring cardio-pulmonary by-pass, a positive TAOT together with a lower A/B threshold of 0.9 could be considered. We have recently used the TAOT alone in serial examination of 188 patients being considered for cardiac surgery, 108 with ischemic heart disease (mean age 54 $\pm$ 8 years) and 80 patients admitted for valve replacement or congenital heart abnormalities (52 $\pm$ 12 years). In 5 of the first group, severe disease was demonstrated and these patients proceeded to carotid arteriography and endarterectomy before the planned heart operation. None of the patients in the second group were found to have severe carotid artery lesions by the ultrasound test (Lewis et al, 1980).

8.5.3   Images

The degree of stenosis measured on the ultrasound images appears to compare well with most of those measured on the X-ray images. However, such comparisons are of limited value since the views are unlikely to be from identical projections. In addition the ultrasound image takes longer to produce and so is more influenced by movement of the wall. The advantage that the simultaneous image and sonogram display give is that changes in blood-velocity can be clearly related to a specific artery or part of an artery. We have not yet attempted to quantify this in our screening but it is a great help in overall assessment of the patient. It is also helpful for demonstrating, by detection of turbulent flow, in which artery a bruit originates.

8.6   GENERAL CONCLUSIONS

It would appear that spectrally analyzed Doppler-shift angiology is
safe and relatively easy to perform. It can be used as an out-patient
procedure, has no contra-indications and the costs are about ten per
cent of those for arteriography. In so far as the measurement of
Doppler-shift is directly consequent on the hemodynamics, the information is complementary to the anatomical vessel geometry given by
X-rays. A great deal of development work has been done on ultrasound
instruments utilizing the Doppler principle since the early publications in the 1960's of such investigators as Satomura and Kaneka
(1960) and Strandness and co-investigators (1966). Doppler blood-
velocimeters with very good directional resolution are now commercially available from several companies as are spectral analyzers,
and it is likely that these instruments - with little change - will
be increasingly used for management of patients.

When the Doppler-shift signal is used in an imaging system then
both vessel geometry and hemodynamic data are available. Since the
frequency and amplitude of the scattered ultrasound can be obtained
in digital form the information can be computer processed so that it
is now theoretically possible to give a more complete description of
the carotid bifurcation than that obtained by angiography. The extent to which this approach, which is at present in the research
stage, is developed over the next decade will be very interesting
to follow.

REFERENCES

Baskett, J.J., Beasley, M.G., Murphy, G.J., Hyams, D.E., and
    Gosling, R.G. (1977). Screening for carotid junction disease by
    spectral analysis of Doppler signals. Cardiovasc. Res. 11, 147-155.

Brockenbrough, E.C. (1970). Screening for the prevention of stroke:
    Use of Doppler flowmeter. Pamphlet. Information and Education Re-
    source Support Unit of the Washington/Alaska Regional Medical
    Program.

Chapman, B., Naylor, G.P., and Charlesworth, D. (1980). Frequency
    analysis of sounds recorded by phonoangiograph. In Diagnosis and
    monitoring in arterial surgery (Edited by R.N. Baird and J.P.

Woodcock). pp. 69-75, John Wright & Son, U.K.

Coghlan, B.A. (1979). Theory and application of spectral analysis
to Doppler-shift ultrasound signals from blood vessels. Ph.D.
Thesis, University of London, London.

Coghlan, B.A. and Taylor, M.G. (1978a). A carotid imaging system
utilizing continuous wave Doppler-shift ultrasound and real-time
spectral analysis. Med. Biol. Eng. Comput. 16, 739-744.

Coghlan, B.A. and Taylor, M.G. (1978b). On methods for preprocessing
directional Doppler signals to allow display of directional blood-
velocity waveforms by spectrum analysers. Med. Biol. Eng. Comput.
16, 549-553.

Coghlan, B.A. and Taylor, M.G. (1979). Improved real-time spectrum
analyser for Doppler-shift blood-velocity waveforms. Med. Biol.
Eng. Comput. 17, 316-322.

Coghlan, B.A., Taylor, M.G., and King, D.H. (1974). On line display
of Doppler-shift spectra by a new time compression analyser. In Cardiovascular applications of ultrasound (Edited by R.S.Reneman) pp.
55-66, North-Holland/American Elsevier, Amsterdam-London-New York.

Duncan, G.W., Gruber, J.O., Dewey, C.F., Myers, G.S., and Lees, R.S.
(1975). Evaluation of carotid stenosis by phono-angiography.
N. Eng. J. Med. 293, 1124-1128.

Edwards, J.H., Kricheff, I.I., Riles, M.D., and Imparato, A. (1979).
Angiographically undetected ulceration of the carotid bifurcation
as a cause of embolic stroke. Radiology 32, 369-373.

Fields, W.S., North, R.R., Hass, W.K., Galbraith, J.G., Wylie, E.J.,
Ratinov, G., Burns, M.H., MacDonald, M.C., and Meyers, J.S.
(1968). Joint study of extracranial arterial occlusion as a cause
of stroke. J.A.M.A. 203, 955-960.

Gee, W., Oller, D.W., and Wylie, E.J. (1976). Non-invasive diagnosis
of carotid occlusion by ocular pneumoplethysmography. Stroke 7,
18-21.

Gomensoro, J.B., Maslenikov, V., Azambuja, N., Fields, W.S., and
Lemak, N.A. (1973). Joint study of extracranial arterial occlusion:
VIII. Clinical-radiographic correlation of carotid bifurcation
lesions in 177 patients with transient cerebral ischemic attacks.
J.A.M.A. 224, 985-991.

Gross, W.S., Verta, M.J. Jr., Bellen, B. van, Bergan, J.J., and
Yao, J.S.T. (1977). Comparison of non-invasive diagnosis techniques
in carotid artery occlusive disease. Surgery 82, 271-278.

Kartchner, M.M., McRae, L.P., Crain, V., and Whitaker, B. (1976). Ocu-
loplethysmography: An adjunct to arteriography in the diagnosis of
extracranial carotid occlusive disease. Am. J. Surg. 132, 728-732.

Lewis, R.R., Beasley, M.G., Ayoub, A., Deverall, P.B., Yates, A.K., and Gosling, R.G. (1980). Diagnosis by ultrasound of severe carotid artery disease in patients undergoing cardiopulmonary bypass operations. Br. Heart J. 43, 414-418.

Lewis, R.R., Beasley, M.G., Hyams, D.E., and Gosling, R.G. (1978). Imaging the carotid bifurcation using continuous wave Doppler-shift ultrasound and spectral analysis. Stroke 9, 465-471.

Lunt, M. (1975). Accuracy and limitations of ultrasonic Doppler blood velocimeter with zero-crossing detector. Ultrasound Med. Biol. 2, 1-10.

Machleder, H.I. (1973). Evaluation of patients with cerebrovascular disease using the Doppler ophthalmic test. Angiology 24, 374-381.

McNeil, B.J., Keller, D., and Adelstein, S.J. (1975). Primer on certain elements of medical decision making. N. Eng. J. Med. 293, 211-215.

Moniz, E. (1927). L'encéphalographie arterielle: son importance dans la localisation des tumeurs cérébrales. Rev. Neurol. 34, 72-90.

Nimura, Y., Matsuo, H., Hayashi, T., Kitabatake, A., Mochizuki, S., Sakakibara, H., Kara, K., and Abe, H. (1974). Studies on arterial flow patterns - instantaneous velocity spectrums and their phasic changes - with a directional Doppler technique. Br. Heart J. 36, 899-907.

O'Donnell, T.F., Pauker, S.L., Callow, A.D., Kelly, J.J., McBride, K.J., and Korwin, O. (1980). The relative value of carotid non-invasive testing as determined by receiver operator characteristic curves. Surgery 87, 9-19.

Reneman, R.S. and Spencer, M.P. (1979). Local Doppler audio spectra in normal and stenosed carotid arteries in man. Ultrasound Med. Biol. 5, 1-11.

Sainz, A., Roberts, V.C., and Pinardi, G. (1976). Phase-locked loop techniques applied to ultrasonic Doppler signal processing. Ultrasonics 14, 128-132.

Satomura, S. and Kaneka, Z. (1960). Ultrasonic blood rheograph. Proc. 3rd Intern. Conf. Med. Electronics. London, 254.

Shapiro, H.M., Lawrence, N.G., Mishkin, M., and Reivich, M. (1970) Direct thermometry, ophthalmodynamometry, auscultation and palpitation in extracranial cerebrovascular disease: an evaluation of rapid diagnostic methods. Stroke 1, 205-218.

Smallwood, R.H. and Brown, B.H. (1978). A measure of blood flow from a frequency analyzed ultrasonic Doppler signal. J. Med. Eng. Technol. 2, 73-74.

Strandness, D.E., McCutcheon, E.P., and Rushmer, R.F. (1966).
Application of a transcutaneous Doppler flowmeter in evaluation of
occlusive arterial disease. Surg. Gynecol. Obstet. 122, 1039-1045.

Wass, W.K., Fields, W.S., North, R.R., Kricheff, I.I., Chase, N.E.,
and Bauer, R.B. (1968). Joint study of extracranial occlusion. II.
Arteriography, techniques, sites and complications. J.A.M.A. 203,
961-968.

Wood, E.H. (1965). Thermography in the diagnosis of cerebrovascular
disease. Radiology 85, 270-283.

CHAPTER 9

The Duplex Scanner—
Real Time Imaging Combined
with Pulsed Doppler

G. Fell *and* D. E. Strandness

9.1 INTRODUCTION

Ultrasonic techniques have been developed to image the carotid artery
bifurcation using continuous wave (Reid and Spencer, 1972), pulsed
Doppler (Mozersky et al, 1971), and pulsed echo (Mercier et al, 1978;
Cooperberg et al, 1979; Leopold, 1978) methods. The carotid artery
bifurcation is an attractive site for noninvasive investigation for
the following reasons:
- atherosclerosis of this segment is responsible for the majority
  (40%) of ischemic strokes
- most atheromatous lesions lie within 2 centimeters of the carotid
  bifurcation
- its relatively superficial location makes it accessible to ultrasound
- atherosclerotic plaques alter the vessel luminal surface and geometry, thereby producing changes in flow characteristics
- the normal flow patterns of the two branches of the common carotid
  artery are quite distinct and it is a relatively simple matter to
  identify the internal and external carotid arteries.

The development of the Duplex Scanner which combines real time
B-mode imaging with pulsed Doppler came about because of the recognition of the limitations noted with continuous wave, pulsed Doppler
and B-mode imaging systems. Calcification may occur either in the wall
of the carotid artery without associated intimal irregularity, or
within the atherosclerotic plaque causing some degree of stenosis.

When this occurs, acoustic impedance is greatly enhanced and the ultrasound is nearly completely reflected, or absorbed, resulting in a blank area of acoustical shadowing beyond the calcific plaque (Hartley and Strandness, 1969). When this occurs, it is difficult to interpret the state of the vessel which is in a sense, hidden by the calcific deposit. Also, it is recognized that some plaques and thrombus have acoustical properties similar to blood and may be missed with B-mode imaging alone. Even totally occluded vessels may appear patent on B-mode ultrasound.

The use of the B-mode image as a guide for precise placement of the sample volume of the pulsed Doppler within any area of interest in the scan plane allows characterization of velocity patterns across the bifurcation. The Duplex scan, when combined with spectral analysis of the Doppler signal, has been found to be accurate in estimating the degree of stenosis and in detecting total occlusion of both the common and internal carotid arteries.

9.2   INSTRUMENTATION

A number of different transducer designs have been devised to produce a real time two-dimensional B-mode image by the pulsed echo technique. Early scanners manually moved the transducer over the skin, but now the ultrasonic transducers are moved mechanically (rotating, translating or oscillating) or a fixed transducer reflects ultrasound off an oscillating acoustic mirror or movement is simulated electronically (linear or annular array). Currently, the best images are produced by the mechanical scanners since they allow the use of large aperture transducers which are more easily focused. A further design constraint on the scan head is noted because of the varying anatomic course of, and desire to obtain access to, the distal extracranial internal carotid artery. The subclavian and vertebral arteries are also of interest and because of their position in the neck, a small, light and highly maneuverable scan head is preferable.

The Duplex concept requires precise registration of the pulsed Doppler sample volume within the visualized image (Fig. 9.1). A number of different methods have been devised to align the pulsed

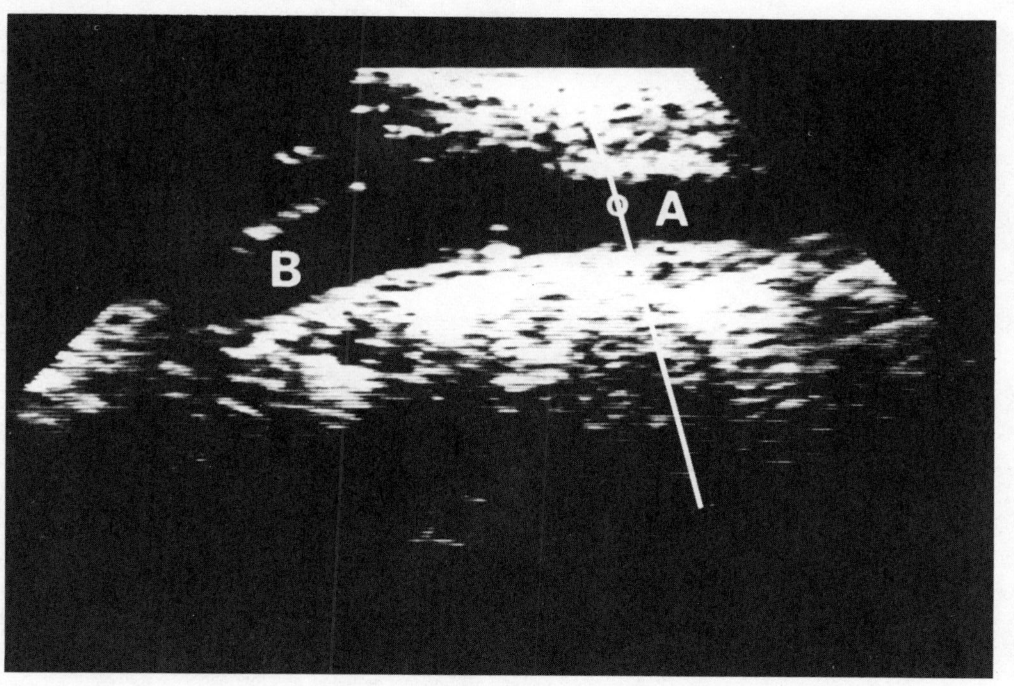

FIG. 9.1   B-mode image of the carotid bifurcation with the patients head to
the right. The position of the Doppler beam is indicated by the white
line with the sample volume represented by the white dot in the internal
carotid artery (A). The common carotid artery (B) but not the external
carotid artery is visualized.

Doppler to the plane of the B-mode image. The initial method
(Philips et al, 1980) placed the Doppler transducer adjacent to the
B-mode transducers, both encased within a water filled silastic boot.
Maintenance of transducer alignment was a serious problem with this
configuration. A more recent solution to the problem of misalignment
inherent in this scheme utilizes the same transducers for both the
B-mode and pulsed Doppler. In one system, the image may be frozen on
a cathode ray tube using a digital field store technique while the
pulsed Doppler beam and sample volume are moved throughout the scan
plane.

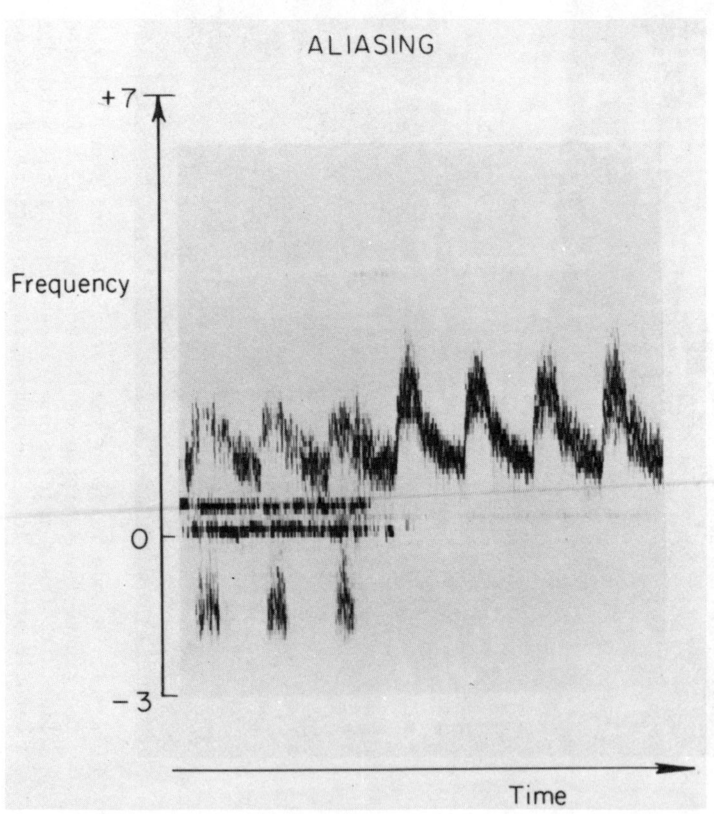

FIG. 9.2   Aliasing of the Doppler signal with a forward flow component appearing in the reverse flow
region. Increasing the sampling frequency (P.R.F.)
correctly displays the Doppler signal.

The pulse repetition frequency (P.R.F.) or sampling rate of any
pulsed Doppler system must be at least twice the frequency of the
expected Doppler shifted frequencies to avoid problems of aliasing
(Fig. 9.2) in which forward flow may appear as reverse flow. Using
5 MHz transducers, one can expect Doppler shifted frequencies well in
excess of 7 kHz in diseased internal carotid arteries where the peak
blood velocity may exceed 250 centimeters per second (Blackshear
et al, 1980). As sound travels at approximately 1500 meters per
second in tissue, a round trip to a depth of 3 centimeters (6 centimeter flight path) takes approximately 40 microseconds. The maximum
theoretical P.R.F. is therefore 25 kHz at 3 centimeters and at a
depth of 5 centimeters the maximum P.R.F. is about 15 kHz. In most
patients an aliased signal should not prove to be a problem because
the majority of carotid arteries are between 2 and 4 centimeters
beneath the skin surface and the peak frequency is usually less than
3.5 kHz, for a 5 MHz transducer, in normal carotid arteries.

The current Duplex system in use at the University of Washington
for carotid artery evaluation consists of several integral parts,
each of which is essential for optimal performance. They include the
following:
- a scan head containing the 5 MHz B-mode and Doppler transducers
  suitable for access to the distal internal carotid artery
- digital field store electronics for high quality B-mode images
- a pulsed Doppler system with quadrature output (Nippa et al, 1975),
  an adequate signal-to-noise ratio and a suitable, small, sample
  volume (approximately 1 cubic millimeter) when used in the range
  of interest
- the ability to freeze frame the B-mode image and to obtain a hard
  copy output of the image, the site and angle of the incident
  Doppler beam from which velocity measurements are made
- fast Fourier transform (F.F.T.) (Cooley and Tukey, 1965) spectrum
  analyzer which permits display of the frequency, amplitude and
  directional characteristics of the audible Doppler signal.

The F.F.T. in use is a digital on-line real time F.F.T. providing
400 spectra per second with a total frequency display of 10 kHz and
a frequency resolution of 100 Hz. Seven kHz are reserved for forward

and 3 kHz for reverse velocity components to the axis of the Doppler beam. Since blood velocity components may be either "toward the transducer or "away from the transducer", depending on vessel anatomy and scan head orientation, the user can arbitrarily select the "direction" to be displayed. Hard copy output of the F.F.T. is obtained on light sensitive paper.

9.3 STUDY PROCEDURE

The Duplex scan is performed with the patient supine and the head supported by a small ring, though it is possible to have the patient elevated 10-15 degrees if they are unable to lie flat. The transducer is coupled to the skin by a water soluble gel and both longitudinal and transverse scans of the carotid arteries may be performed, depending upon the scan head orientation. However, scans in the longitudinal plane have been found to be the most useful. The vessels are readily identified on the B-mode image by the bright wall echoes separated by the dark echo free vessel lumen and confirmed by the characteristic audible Doppler signal obtained from the center of the vessel (Fig. 9.1). The internal and external carotid artery flow often goes to zero or reverses during diastole due to the higher peripheral resistance of the vascular beds it supplies (Fig. 9.4E). The B-mode images are inspected for acoustical shadowing or bright echoes from plaques and the audible Doppler signals analyzed for both the presence of high frequencies and harsh multiple or broad frequencies associated with turbulence. Calcification is usually confined to the region of the carotid bifurcation and a Doppler signal can be readily obtained from a more distal site in the internal carotid artery where flow disturbances originating from lesions in the shadowed region can be detected.

This study is begun low in the neck with the transducer against the clavicle, the head turned slightly toward the opposite side and the neck extended. The common carotid artery is initially identified and moving cephalad the carotid bulb is recognized as a slight dilatation of the artery. The bifurcation is rarely visualized in one plane and the internal carotid artery is usually located postero-laterally to

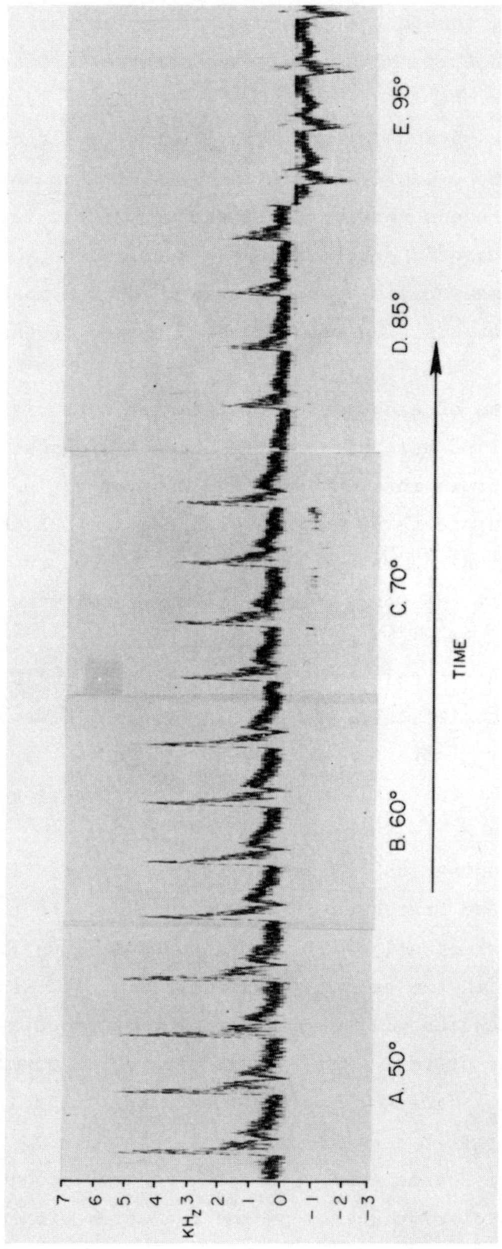

FIG. 9.3   Spectrum from the common carotid artery demonstrating the effect of the angle of the incident Doppler beam on the peak frequency and direction of the Doppler shifted signal.

the external carotid artery. The anatomic variations of the bifurcation are usually slight, though the two branches may be tortuous or
lie very close to one another and this may necessitate turning the
patient's head to one or the other side to obtain precise identification by scanning through various planes. The internal and then external carotid arteries can be easily identified by listening to their
audible Doppler signals and observing their characteristic velocity
patterns. The internal carotid artery may be visualized to the angle
of the mandible and in many cases beyond the angle of the mandible by
scanning from a posterior position with the head turned to the opposite side.

Precise identification of each vessel of interest both visually
and audibly is essential to help establish a correct diagnosis. A
potential source of error is introduced if the Doppler angle is not
kept constant, usually close to 60 degrees with the axis of the vessel. Fig. 9.3 demonstrates the marked change in peak frequency and
even direction of flow if the Doppler angle is neglected when recording the pulsed Doppler signal.

High grade stenoses and total occlusions can usually be identified
from the audible characteristics of the Doppler signal; however, flow
disturbances produced by minor stenoses may go undetected by the ear.
The addition of real time spectral analysis enhances the capability
of this system to detect minimal stenoses.

At present, the demonstration of a stenosis or occlusion is most
reliably determined by the frequency distribution seen with the spectrum analyzer (Blackshear et al, 1979). In the normal situation in
which all the red cells in the center stream move with the same velocity, the band of frequencies will remain narrow throughout systole
and the area beneath the systolic envelope is clear. The normal
spectrum consists of peak frequencies less than 3 kHz, usually 2-3
kHz, and a clear area beneath the systolic peak (Fig. 9.4A).

Increasing stenosis is associated with both spectral broadening
and increasing peak systolic frequency as the red cells within the
pulsed Doppler sample volume move in a random pattern and with increased linear velocity. With minimal flow disturbance, spectral
broadening occurs initially during the deceleration phase of systole

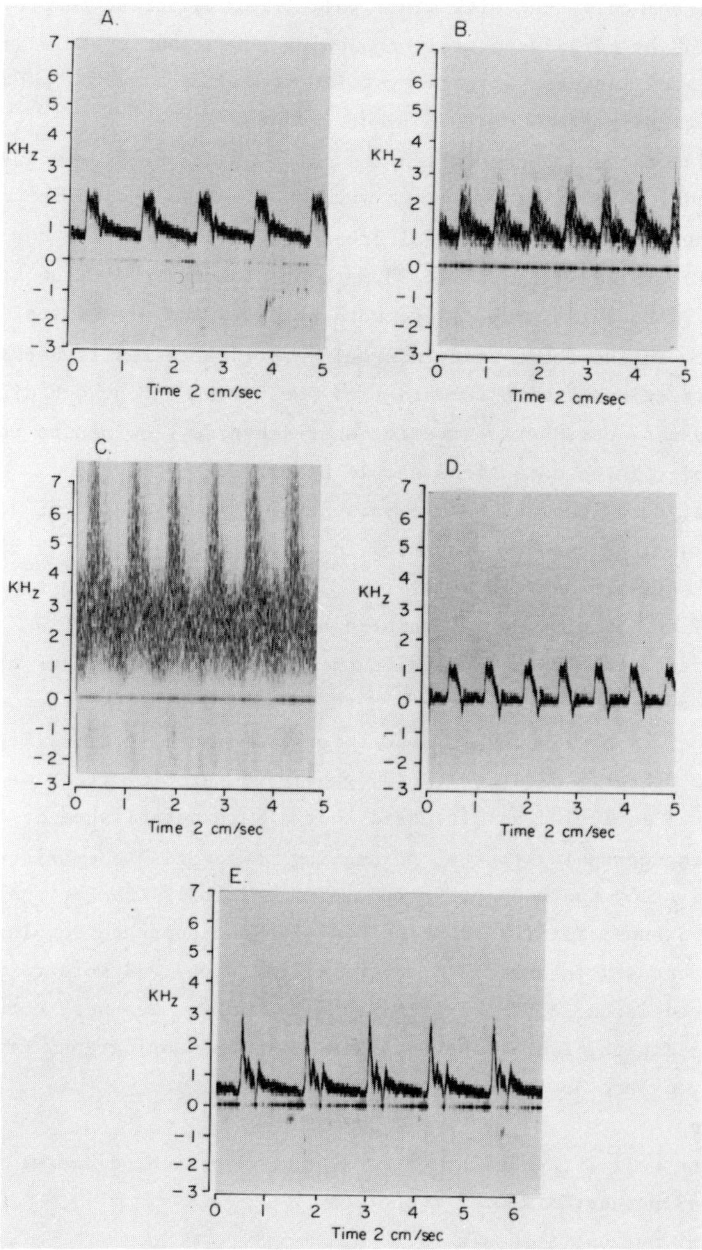

FIG. 9.4   Spectrum of carotid arteries
with varying grades of stenosis
(see text).

and is commonly seen with minor wall irregularities. Complete "fil-
ling in" or marked spectral broadening throughout systole without an
associated increase in peak systolic frequency above 3.5 kHz is seen
with stenosis of 10 to 40 percent diameter reduction (Fig. 9.4B);
that is, those stenoses which are not pressure or flow reducing. Peak
frequencies in excess of 4 kHz with an elevated diastolic velocity
component and marked spectral broadening throughout the cardiac cycle
are associated with a 50 to 99 percent diameter reduction stenosis
(Fig. 9.4C) which will reduce both pressure and flow distally.

Flow in the common and internal carotid arteries is normally
"unidirectional" with forward flow persisting throughout diastole in
contrast to peripheral arteries where reverse flow occurs in early
diastole. Three exceptions should be noted:
- some young, healthy normal patients with very compliant vessels
  may demonstrate flow reversal in early diastole (Fig. 9.5)
- patients with aortic valve incompetence may have flow reversal
  bilaterally with normal carotid bifurcations
- patients with total occlusion or high grade stenoses of the internal carotid artery may have a short period of flow reversal in
  association with low or absent forward diastolic flow (Fig. 9.4D).

The criteria used for the diagnosis of carotid artery disease by
spectral analysis (as discussed above) were established by comparative angiographic studies. An ongoing prospective comparison-and-va-
lidate study has been performed using initially the Peripheral Vascular Scanner III (PVIII -- pulsed echo and separate Doppler transducer encased in a silastic water-filled boot) and more recently the
Advanced Technology Laboratories Mark V Duplex Scanner, both combined
with a digital F.F.T. spectral analyzer. The angiography criteria
employed were as follows:
- normal
- minor wall irregularity or less than 10 percent diameter reduction
- 10-49 percent diameter reduction
- 50-99 percent stenosis
- total occlusion.

Duplex studies were performed in over 640 patients using the PVIII
Scanner with 183 carotid bifurcations studied by angiography

available for comparative study. The ability to detect carotid artery disease or sensitivity by Duplex scanning was 97%, 177/183.

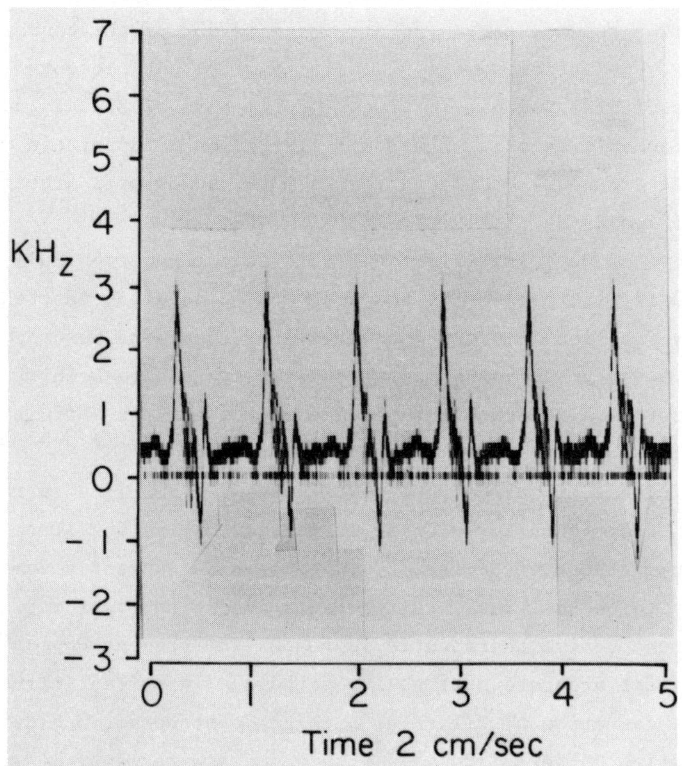

FIG. 9.5   Spectrum from the common carotid artery
of a young patient demonstrating flow reversal
due to compliant vessels.

When an attempt was made to separate the patients into the angiographic categories based on the peak frequencies and degree of spectral broadening noted on spectral analysis, the sensitivity for each category was as follows:

- less than 10% stenosis: 3/9,   33%

- 10-49% stenosis: 40/79, 51%

- 50-99% stenosis: 57/62, 92%

- total occlusion: 28/33, 85%

    Thus, while it might appear that accurate angiographic grading by

the velocity changes is not nearly as good as detecting disease per
se, it must be remembered that the inter- and intraobserver variability on the part of angiography is also not very good. For example,
when the same radiologist reread the arteriogram using the above
classification scheme, there was agreement in 83% of the cases. In
other words, the arteries were placed in a different category in 17%
of the cases (Chikos et al, in press; Croft et al, 1980). In addition,
the angiographic diagnosis of carotid ulceration is uncertain with
an incorrect diagnosis of ulceration in 17 of 50 carotid arteries in
one study (Edwards et al, 1979).

The number of Duplex studies available with angiographically
normal carotid bifurcations is too small to be able to state the
specificity (ability to predict normalcy) of the test. However, large
numbers of young presumed normal carotid arteries have been studied
and they provide a reasonable reference source for the definition of
the criteria, characteristic for a normal study.

The usefulness of current Duplex scanning systems lies in the
evaluation of patients known to be at increased risk for stroke
rather than in those with classic symptoms which currently demand
angiography to be the final arbiter in the decision regarding a surgical or conservative therapeutic approach. The present Duplex system is the most accurate noninvasive technique in current clinical
use for the detection of bilateral high-grade stenosis, occlusion and
differentiation of normal vessels from vessels with minor degrees of
stenosis. The major drawbacks of the Duplex system are the inability
to either visualize on the B-mode image or predict by Doppler signal
processing techniques the presence of carotid artery ulceration, the
identification of intracranial carotid siphon or aortic arch lesions.

Detection and diagnosis of other less common problems involving
the extracranial cerebral vascular system is also possible with
Duplex scanning. This includes the imaging of and detection of the
velocity changes associated with tortuous, kinked or aneurysmal
carotid arteries (Fig. 9.6). Arteriovenous fistulae involving the
carotid arteries due to intracranial congenital abnormalities
(Fig. 9.7) or as a result of cervical or cranial trauma giving rise
to symptoms or carotid bruits can be diagnosed by the velocity

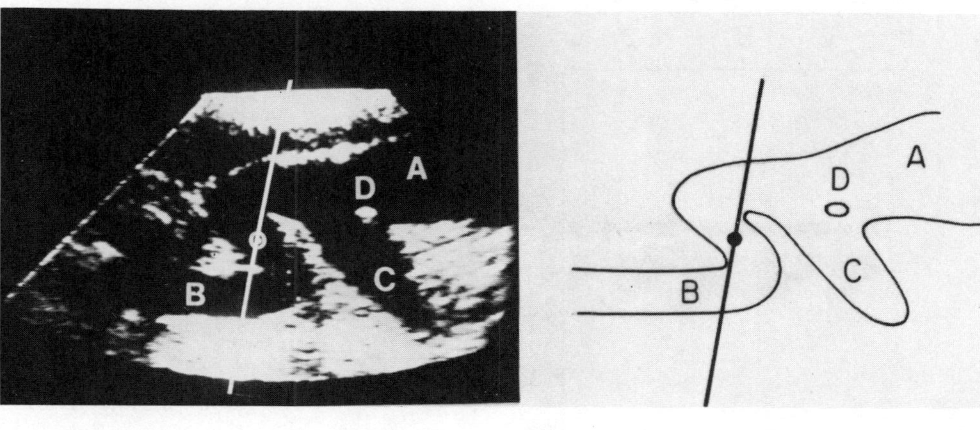

FIG. 9.6   B-mode image of suspected carotid artery aneurysm. The innominate (A), common carotid (B) and right subclavian (C) arteries are all visualized. A calcified plaque (D) producing an area of acoustical shadowing is noted at the origin of the subclavian artery.

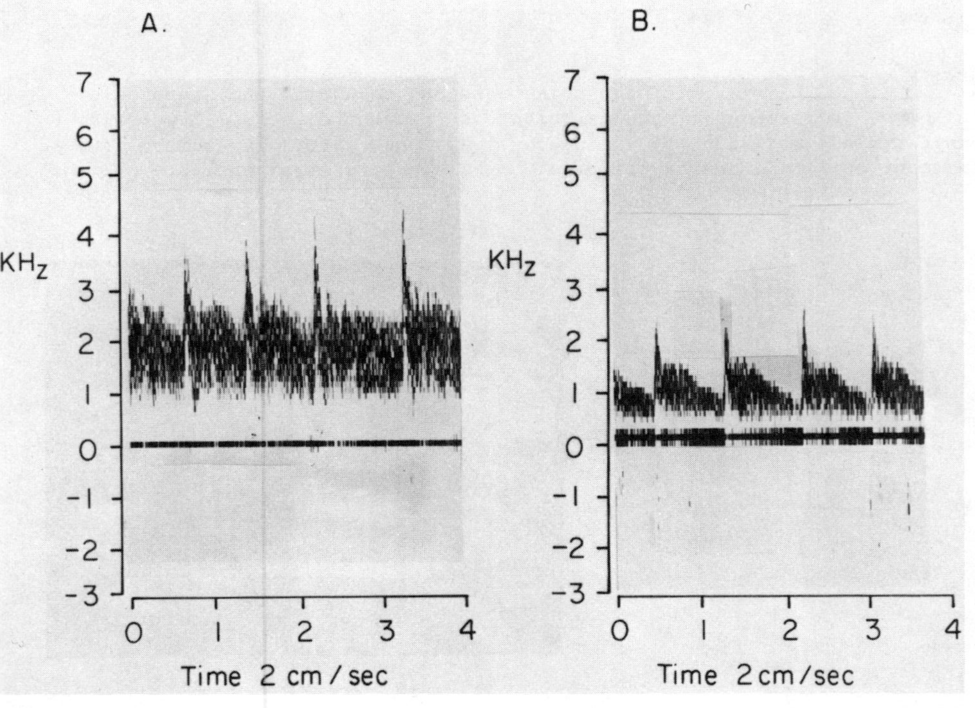

FIG. 9.7   Spectrum from the internal carotid arteries of a young child. High turbulent diastolic velocities (A) in the left internal carotid artery are in marked contrast to the right internal carotid artery, suggesting the presence of an arteriovenous malformation.

changes noted on spectral analysis of the Doppler signal. Subclavian
artery disease, stenotic, aneurysmal or steal syndromes can also be
investigated by Duplex techniques, though accurate anatomical
definition and diagnosis is difficult.

9.4 CONCLUSIONS

The long-term aim of Duplex scanning must be to develop methods
of improved vessel imaging in conjunction with improved quantitative
rather than qualitative signal processing which can accurately predict those truly at risk for stroke. However, this new technology is
already providing pathophysiological information which will require
long-term model and animal experimentation in conjunction with
clinical investigation to fully understand the hemodynamic mechanisms
involved in the production of the ischemic stroke syndrome.

REFERENCES

Blackshear, W.M., Phillips, D.J., Chikos, P.M., Harkey, J.D.,
    Thiele, B.L., and Strandness, D.E.,Jr. (1980). Carotid artery
    velocity patterns in normal and stenotic vessels. Stroke 11, 67-71.

Blackshear, W.M. Jr., Phillips, D.J., Thiele, B.L., Hirsch, J.H.,
    Chikos, P.M., Marinelli, M.R., Ward, K.J., and Strandness, D.E. Jr.
    (1979). Detection of carotid occlusive disease by ultrasonic imag-
    ing and pulsed Doppler spectrum analysis. Surgery, 86, 698-706.

Chikos, P.M., Fischer, L., Hirsch, J.H., Harley, J.D., Thiele, B.L.,
    and Strandness, D.E., Jr. (in press). Observer variability in
    evaluating extracranial carotid artery stenosis. Am. J. Neuroradiol.
    In press.

Cooley, J.W. and Tukey, J.W. (1965). An algorithm for the machine
    calculation of complex Fourier series. Math. Comp. 19, 297-301.

Cooperberg, P.L., Robertson, W.D., Fry, P., and Sweeney, V. (1979).
    High resolution real time ultrasound of the carotid bifurcation.
    J. Clin. Ultrasound 7, 13-17.

Croft, R.J., Ellam, L.D., and Harrison, M.J.G. (1980). Accuracy of
    carotid angiography in the assessment of atheroma of the internal
    carotid artery. Lancet 1, 997-999.

Edwards, J.H., Kricheff, I.I., Riles, T., and Imparato, A. (1979).
    Angiographically undetected ulceration of the carotid bifurcation

as a cause of embolic stroke. Radiology, 132, 369–373.

Hartley, C.J. and Strandness, D.E., Jr. (1969). The effects of atherosclerosis on the transmission of ultrasound. J. Surg. Res. 9, 575–582.

Leopold, A.R. (1978). Pulse echo ultrasonography. In Non-invasive diagnostic techniques in vascular disease (Edited by E.F. Bernstein). pp. 23–28, The C.V. Mosby Co. St. Louis, Mo.

Mercier, L.A., Greenleaf, J.F., Evans, T.L., Sandok, B.A., and Haltery. (1978). High resolution ultrasound arteriography: A comparison with carotid angiography. In Non-invasive diagnostic techniques in vascular disease (Edited by E.F. Bernstein). pp. 231–244, The C.V. Mosby Co., St. Louis, Mo.

Mozersky, D.J., Hokanson, D.E., Baker, D.W., Sumner, D.S., and Strandness, D.E., Jr. (1971). Ultrasonic arteriography. Arch. Surg. 103, 663–667.

Nippa, J.H., Hokanson, D.E., Lee, D.R., Sumner, D.S., and Strandness, D.E., Jr. (1975). Phase rotation for separating forward and reverse blood flow velocity signals. IEEE Transactions on Sonics and Ultrasonics SU-22, 340–346.

Phillips, D.J., Powers, J.E., Eyer, M.K., Blackshear, W.M., Jr., Bodily, K.C., Strandness, D.E., Jr., and Baker, D.W. (1980). Detection of peripheral vascular disease using the duplex scanner III. Ultrasound Med. Biol. 6, 205–218.

Reid, J.M. and Spencer, M.P. (1972). Ultrasonic Doppler technique for imaging blood vessels. Science 176, 1235–1236.

ACKNOWLEDGEMENT

This work was supported by National Institutes of Health Grant ≠ HL 20898-03.

CHAPTER 10
Prospects and Conclusions
R. S. Reneman, L. Pourcelot *and* A. P. G. Hoeks

10.1   INTRODUCTION

In the clinically oriented chapters of this book several techniques
have been described to diagnose cerebrovascular disease and to evaluate the functional state of the cerebral circulation with Doppler
instruments. With most of these techniques vascular lesions associated with substantial narrowing of the carotid artery (more than
50-60% decrease in diameter) can be detected accurately, while they
can distinguish between tight stenosis (more than 90% decrease in
diameter) and total occlusion. The latter is important because in
tight stenosis high shear stresses are considered to induce thrombus
formation causing, for example, Transient Ischemic Attacks (TIA's)
due to emboli. Accurate diagnosis of lesser degrees of stenosis, however, is still problematic, although promising results have been obtained with shape analysis of the audio spectrum (see Chapter 8) and
by detecting disturbances in the flow pattern (see Chapter 9). These
disturbances do occur at relatively slight degrees of artery narrowing (Giddens et al, 1976; Barnes et al, 1976; Sandmann et al, 1978).

The detection of lesions in the internal carotid artery without or
with slight narrowing of the artery, is important because they are
believed to cause severe cerebral disturbances through emboli
(e.g. TIA's). Moreover, the diagnosis of vascular lesions at an early
stage of the disease will be helpful in epidemiological studies as
well as in obtaining more insight in the natural course of the disease.

At the present state of the art in the clinic, disturbances in the flow pattern are generally diagnosed by estimating the degree of broadening of audio spectra. In spite of the promising clinical results, this method has its limitations. Spectral broadening does not necessarily mean pathology, while from audio spectra no information can be derived about the velocity profile - that is the velocity distribution over the cross-sectional area of the blood vessel - at discrete time intervals during one cardiac cycle. These profiles, which were found to change locally in the vicinity of a stenosis (Wille, 1979), can be recorded on-line with multi-channel pulsed Doppler systems (Peronneau et al, 1974; Anliker, 1978; Brandestini, 1978; Hoeks et al, 1981). Proper localization of the disturbance in the flow pattern can be achieved by combining the Doppler device with velocity imaging or B-mode imaging of the arterial wall. The latter technique is also in use to detect atherosclerotic lesions without substantial narrowing of the artery.

In this chapter recent ultrasonic developments that may be of clinical importance will be discussed. Special attention will be paid to the diagnosis of vascular lesions at an early stage of the disease and real-time imaging. Finally, we will try to summarize which of the available Doppler instruments are of use in the clinic to evaluate the cerebrovascular circulation.

10.2   THE USE OF SPECTRAL BROADENING TO DETECT DISTURBANCES IN THE
        FLOW PATTERN

As mentioned above, disturbances in the flow pattern do occur at relatively slight degrees of arterial narrowing so that the detection of these disturbances can be considered as an important parameter in the diagnosis of arterial lesions at an early stage of the disease.

As discussed in section 4.5.2 rather detailed information about the disturbances in the flow pattern along stenosed arteries can be obtained by using a CW Doppler device with velocity imaging and audio spectrum analysis. This method also detects arterial wall vibration as induced by high velocity turbulence just distal to a stenosis. Velocity imaging is used to localize properly the site of recording

of the audio spectrum in relation to the position of the stenosis.
Another approach is to combine audio spectrum analysis with a single-
channel pulsed Doppler system and to localize the site of sampling by
using a B-mode image of the vessel wall (see Chapter 9). Advantages
of this approach are that information can be obtained not only about
the angle between the sound beam and the direction of blood flow, but
also changes, if any, in the vessel wall may be detected (see Section
10.4.2) and contamination of the velocity signal by unwanted signals,
like those from the vessel wall, can largely be prevented. A drawback
of using simultaneously pulsed Doppler and B-mode imaging techniques
is that the maximum detectable velocity, which is already limited in
pulsed Doppler systems (see Section 4.4.2), is further reduced. This
can cause unreliable readings, especially in the vicinity of stenotic
lesions.

Although audio spectrum analysis yields valuable information about
the flow disturbances in stenosed arteries, this method has its limitations. Spectral broadening is generally used to diagnose disturbances in the flow pattern and to estimate the degree of carotid artery narrowing. Spectral broadening, however, does not necessarily mean
pathology and can also occur when in the sample volume:
- the velocity profile is parabolic rather than flat (plug flow)
- the sound beam is non-homogeneous and divergence of the beam occurs.
In CW-systems the dimensions of the sample volume are mainly determined by the width of the ultrasound beam, while in pulsed Doppler
systems the sample volume depends on both the beamwidth and the effective sample duration.

These considerations imply that the width of the audio spectrum
depends on the equipment used so that the spectral broadening due to
the equipment has first to be known before the obtained results can be
interpreted properly. It also implies that the spectrum recorded with
CW-systems is broader than that recorded with pulsed devices
(Fig. 10.1). In the latter devices spectra are displayed of samples
usually taken from the centre of the vessel where red blood cells are
travelling at approximately the same high velocity in case of plug
flow and even parabolic flow. In CW-systems velocities are sampled
along the whole cross-section so that the lower velocities near the

274

vessel wall are presented in the audio spectrum as well.

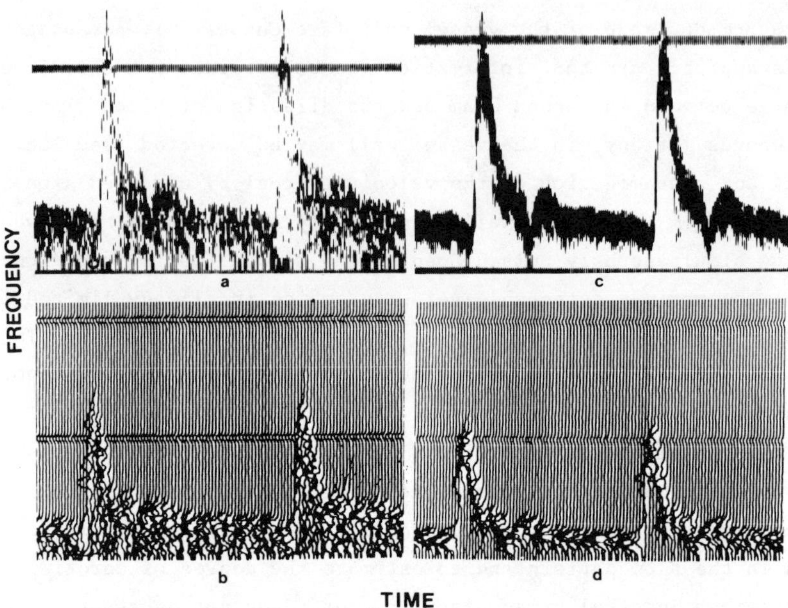

FIG. 10.1   Frequency distribution as a function of time in
     intensity mode (a and c) and amplitude mode (b and d) as
     recorded with CW (a and b) and pulsed Doppler instruments
     (c and d) in the common carotid artery of a healthy subject.

An additional problem encountered in CW devices is that the effect
of beam divergence is more prominent at smaller angles of insonation, especially when focussed transducers are used, causing additional spectral broadening. In CW-systems spectral windows - i.e.
the open area in the spectrum, representing velocities at which
practically no red blood cells are travelling - are not always
clearly visible (Fig. 10.1). Therefore, pulsed Doppler systems are
likely to be more suitable to determine the degree of spectral
broadening than CW-systems (Blackshear et al, 1979). As discussed
in chapter 4 high amplitude low frequencies due to lateral wall
motion may appear in the audio spectrum as a result of sampling over
the full range. This is a disadvantage because the high amplitude of
these low frequencies can mask the presence of high velocity infor-

mation in the spectrum.

Beside the above-mentioned reasons for spectral broadening, frequency dispersions will be introduced by the finite size of the sample volume in the direction of the velocity (transit-time effects), which will be dependent among other factors upon the angle of insonation (α in equation 4.1). The smaller this dimension is, the larger the spectral broadening will be. Therefore, the effect will be more prominent in pulsed Doppler systems where small sample volumes are employed. Actually, the spectral broadening due to the finite sample size is related to the average Doppler frequency (if the observed velocity distribution is flat) which means that this spectral broadening relative to the average frequency is constant for a given system. In absolute terms the transit time effect will become more prominent at higher average Doppler frequencies.

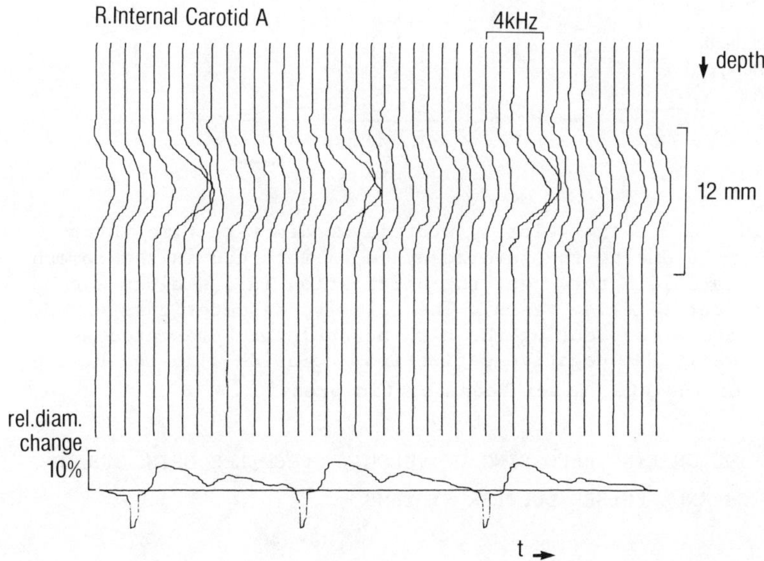

FIG. 10.2   The velocity profile at discrete time intervals
during the cardiac cycle as recorded in the internal carotid artery in a healthy subject (age: 20 years). The
relative diameter changes of the artery during the cardiac cycle are presented as well. The negative deflections coincide with the R-wave of the ECG

One should realize that audio spectra do not yield information about the velocity profile at discrete time intervals during the cardiac cycle. This limits their use in the diagnosis of arterial lesions because in the vicinity of a stenosis the velocity profile was found to change locally during one cardiac cycle (Wille, 1979).

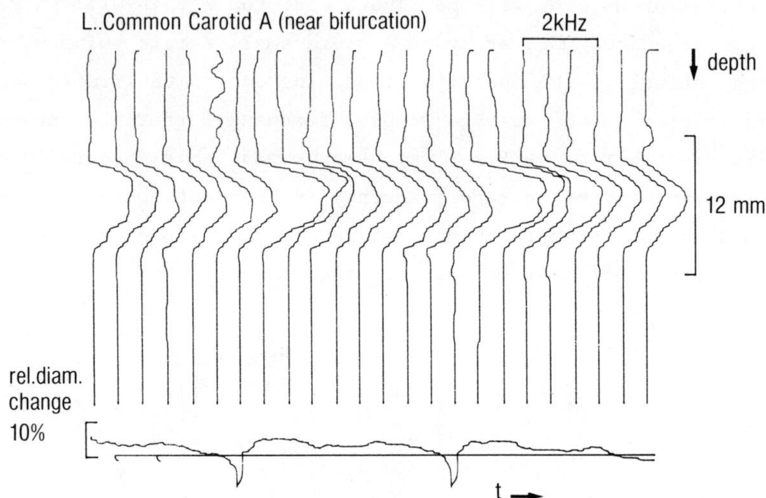

FIG. 10.3   The velocity profile at discrete time intervals during the cardiac cycle as recorded in the common
carotid artery near the bifurcation in a healthy subject (age: 45 years). The relative diameter changes of
the artery during the cardiac cycle are presented as
well. The negative deflections coincide with the R-wave
of the ECG. After Reneman, (in press).

10.3   THE ON-LINE RECORDING OF VELOCITY PROFILES USING MULTI-
CHANNEL PULSED DOPPLER SYSTEMS

Velocity profiles at discrete time intervals during one cardiac cycle can be recorded on-line with multi-channel pulsed Doppler systems (Peronneau et al, 1974; Anliker, 1978; Brandestini, 1978; Hoeks et al, 1981). Besides, these systems allow the on-line recording of the relative diameter changes of an artery during the cardiac cycle (Hoeks et al, 1980; Reneman, in press). In healthy adults the velocity profile as recorded in the common (Fig. 4.7) and internal carotid arteries (Fig. 10.2) is generally symmetric and flat, although in the latter artery often a more parabolic profile is seen at mid-systole. A flat profile is representative of plug flow (see section 4.5.2). Near the carotid bifurcation the velocity profile is usually asymmetric, especially at peak velocity during systole (Fig. 10.3). The relative changes in diameter of the common carotid artery during systole decrease with age (cf Figs. 4.6 and 10.3). Under normal circumstances the anterior and posterior wall are contributing approximately equally to the increase in diameter of this artery as indicated by the symmetric widening of the velocity profile during systole (Fig. 4.7).

There are indications that information about the presence and localization of lesions can be derived from both the velocity profile at discrete time intervals during the cardiac cycle and the mean velocity as an instanteneous function of time as recorded at various sites in an artery (Fig. 4.6). Indicative of vascular lesions are:
- asymmetry of the velocity profile at sites where the profile is symmetric under normal circumstances
- asymmetric widening of the velocity profile during systole
- a narrow velocity profile in combination with high peak velocities during systole
- distortion of the velocity waveforms and oscillations on these tracings locally in the artery or along its cross-section, representing the presence of turbulence.

Proximal to a small stenotic lesion (less than 50% decrease in diameter) in a common carotid artery, the velocity profile was found to be asymmetric due to reflections or local contraction of the blood stream or both (Fig. 10.4). The asymmetry of the velocity profile presented in this figure was most pronounced near the posterior wall, indicating that the lesion was located at this site of the vessel.

Although preliminary clinical findings indicate that changes in the velocity profile at discrete time intervals during the cardiac cycle and the mean velocity as an instantaneous function of time as recorded simultaneously at various sites in an artery, do yield

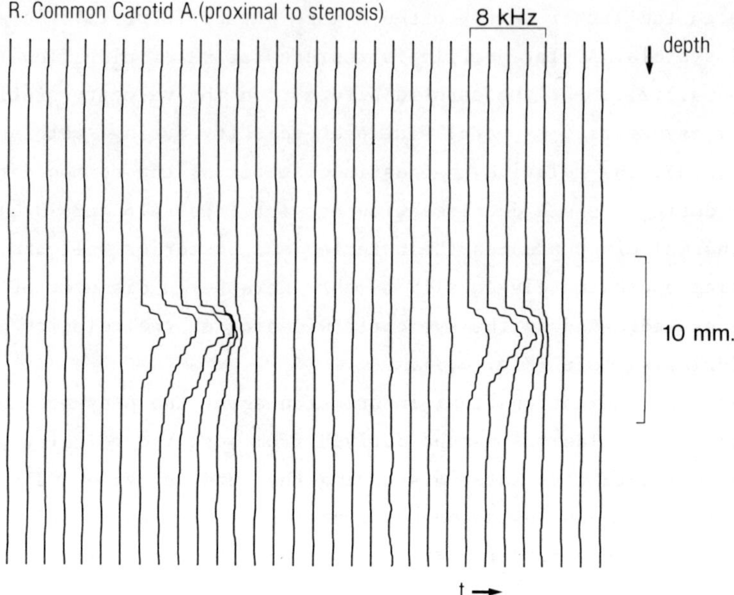

FIG. 10.4   The velocity profile at discrete time
    intervals during the cardiac cycle as recorded in
    the common carotid artery proximal to a small ste-
    notic lesion (<50% narrowing). Note the asymmetry
    of the velocity profile near the posterior wall.

information about the presence and localization of arterial lesions,
several problems are encountered in this approach:

- quantification of the disturbances in the velocity profile is
  difficult
- the maximum velocity that can be detected unambiguously is limited
  (see Section 4.4.2) so that information about lesions has to be
  derived from recordings proximal to and distal to the lesion
- positioning and maintenance of the sample volume within a tight
  stenosis are difficult, especially when narrow sound beams are
  used.

Further clinical investigations are required to be able to decide
whether the on-line recording of velocity profiles and instantaneous
velocity waveforms at various sites in an artery are an asset in the
diagnosis of vascular lesions, especially those associated with

slight narrowing of the artery. Theoretically the resolution of the
present generation of the multichannel pulsed Doppler system as
developed in Maastricht is sufficient to detect local disturbances
in the velocity profile. Preliminary *in vitro* experiments have shown
that 5% narrowing of a tube results in disturbances in the velocity
profile which can be detected with this system. The characteristics
of the Maastricht pulsed Doppler system are presented in Table 10.1.

TABLE 10.1   Characteristics pf the present generation of the
multichannel pulsed Doppler system as developed
in Maastricht

Emission frequency	6 MHz
Pulse repetition frequency	9 or 18 kHz
Duration of emission	0.5 or 1 μs
Sample gate-width	0.5 μs
Sampling distance	0.5 or 1 mm
Number of gates	128[1]
Accuracy of vessel wall displacement measurement	32 μm

1) can be selected in multiples of 8

10.4   VELOCITY IMAGING AND B-MODE IMAGING OF THE VESSEL WALL

10.4.1 <u>Introduction</u>

For proper localization of the disturbance in the flow pattern, the
Doppler instrument should be combined with velocity imaging or B-mode
imaging of the vessel wall. The latter technique may also be used to
detect vascular lesions not associated with substantial narrowing of
the artery. For this purpose velocity and B-mode imaging can be combined (Pourcelot, 1977; Eyer et al, 1981). Since velocity imaging as
such is sufficiently discussed in the chapters 4, 7 and 8, we will
confine ourselves, in the present chapter, to B-mode imaging and the
combination of both imaging techniques.

10.4.2 <u>B-mode Imaging</u>

In general Doppler systems discard the echo information present in
the received signal. The returned echo signal is of poor quality, because of the angle under which the arterial walls are observed. Since
the angle of incidence is equal to the angle of reflection the returned signal will have the largest amplitude if the reflector is perpendicular to the ultrasound beam.

Pulsed echo systems designed to visualize structures are better
suited to pick up echoes reflected by the vascular walls. In B-mode
echo systems the amplitude of the received echo as function of depth
is represented by brightness (B) modulation of the CRT-beam, i.e.
the amplitude is displayed in shades of grey. To make a two-dimensional scan either the direction (sector scanner) or the position
(linear scanner) of the ultrasound beam or both (compound scanner)
should be altered for subsequent pulses. Compound scanners are less
suited for real-time visualization because of the low frame-rate.

Changing the position and/or direction of the ultrasound beam can
be achieved either mechanically or electronically. Mechanical linear
scanners have a large scanhead and hence are difficult to manipulate.
Mechanical sector scanners have a considerable smaller scanhead and
can be used for visualization of the arteries in the neck. Linear
electronic scanners employ a probe consisting of an array of individual transducers. Lateral shifting of the ultrasound beam is accom-
plished by activating the next transducer or a subset of transducers
if dynamic focussing of the ultrasound beam is desired. The probe of
an electronical sector scanner (phased array scanner) consists of a
large number of transducer elements which act as a single transducer.
The direction of the ultrasound beam can be changed by introducing
time-lags between the emitted pulses of the individual elements. The
same procedure has to be performed upon the received signals before
adding them. The probe of an electronic sector scanner is much
smaller than the probe of an electronic linear array scanner. Presently, focussing techniques for electronic scanners result in narrowing of the ultrasound beam in one lateral dimension while the
other dimension is determined by the dimension of the transducer

element. Mechanical scanners employ rotation symmetric transducer(s) allowing axi-symmetric focussing. In both kinds of systems a narrow beamwidth and a good axial resolution can be achieved by utilizing a high emission frequency (5-10 MHz) at the expense of a reduced depth of penetration (Fig. 10.5a)

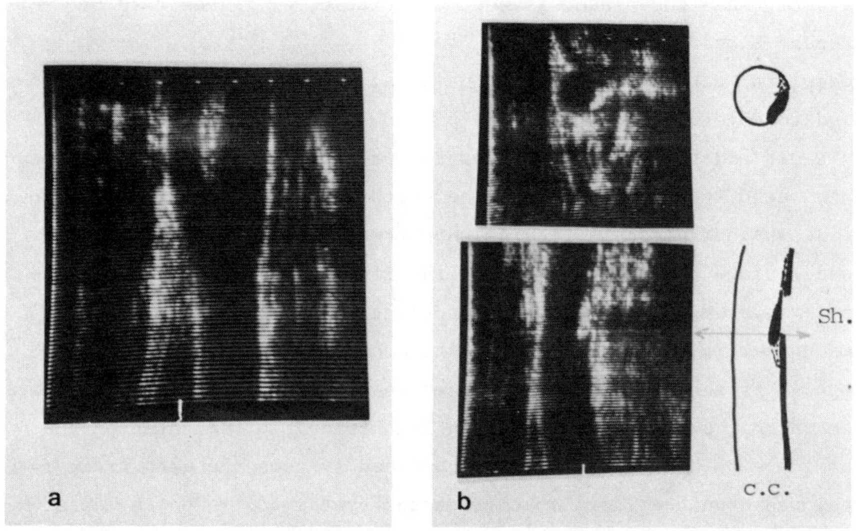

FIG. 10.5   a: Longitudinal section of a normal carotid bifurcation, using a linear array of 100 transducers, dynamically focussed and working at 5 MHz.
b: Transversal and longitudinal section of the common carotid artery in a patient with atheromatous plaques as obtained with a 5 MHz linear array scanner. Notice the shadowing (sh) behind the plaque.

Sector scanners may introduce misleading artefacts in the echo image if structures running parallel to the skin surface are visualized (like the carotid arteries) because of the changing angle of incidence. Misinterpretation can be avoided if a suspected region is investigated from different directions. A small probe facilitates manipulation of the scan plane.

In vascular pathology and especially in cases of atheromatous deposits, the results obtained with real-time imaging are still

difficult to interpret because of the varying echographic responses.
Schematically one distinguishes three types of images:
- clear echographic information with the atheromatous plaque clearly
  visible (Fig. 10.5b)
- a poor response with very few echoes within the plaque; for instance, when the plaque is soft or brittle, or in the case of
  recent blood clots
- complete lack of information due to acoustical shadows often related to calcium deposits in the plaque or in the arterial wall.

   To get additional evidence for the presence of a plaque two-dimensional echo systems can be combined with a pulsed Doppler system to
decide whether blood is flowing unobstructed through the imaged
vessel. Since the echo system may interfere electronically with the
Doppler system, both systems are not activated simultaneously. In
some commercially available combined echo Doppler systems the echo-
image is frozen while switching over to the Doppler system. The place
from which the Doppler information is obtained is indicated on the
echo image by a marker. In other combined systems the echo frame rate
is slowed down considerably to free the system for velocity measurements. Both echo and Doppler information can be assessed by the same
transducer (then the transmitted frequency for both systems is auto-
matically the same) or two transducers can be combined in one single
probe (allowing an independent choice for the emission frequency of
both systems).

   The commercially available combined echo/Doppler systems can assess the Doppler information in a single sample volume. Pulsed echo
systems can also be combined with a multi-channel pulsed Doppler
system. The velocity information, as sensed along the Doppler ultrasound beam, can then be projected on the echo-image (Green et al,
1977). If the echo system is used in M-mode (B-mode as a function of
time along a single line of observation), echo and velocity distribution along the ultrasound beam can be superimposed in a single
image (Eyer et al, 1981). To distinguish echoes from forward and
reversed flow the picture is color coded. Two-dimensional echo and
velocity mapping, utilizing a linear electronic scanner has been
described by Pourcelot (1977). In this system the flow scan can be

displayed separately (Fig. 10.6) or superimposed on the B-mode
(Pourcelot, 1977).

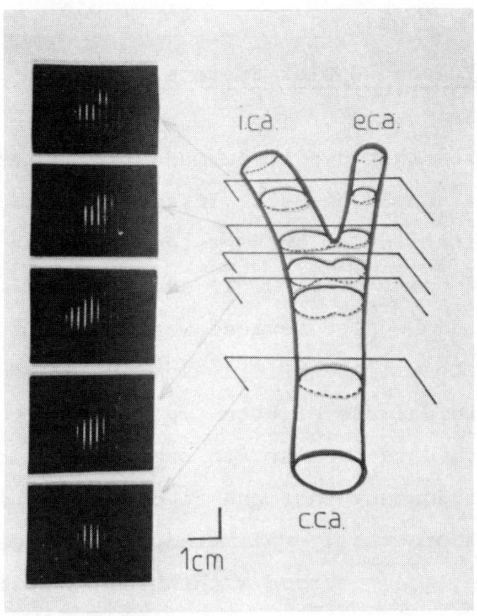

FIG. 10.6   Doppler tomographies
of the carotid arteries, using
a multi-channel pulsed Doppler
system. A rapid translation
of the Doppler probe is
achieved using an array of
10 transducers. Each cross-
section is obtained in 1/15th
of a second.

10.5   WHICH INSTRUMENTS ARE NEEDED FOR ADEQUATE EVALUATION OF
THE CEREBRAL CIRCULATION WITH DOPPLER ULTRASOUND

10.5.1   Introduction

Different techniques to evaluate the cerebral circulation in the
clinic with Doppler ultrasound have been described in this book. In
the past decade these techniques have been developed by experts with
a long history in ultrasound. Therefore, the question can be raised

whether all of the described instruments should be present in a cerebrovascular laboratory of a general hospital. In this section an attempt is made to indicate the possibilities of the various instruments, which may facilitate their selection for clinical use.

10.5.2 CW Versus Pulsed Doppler Systems

An important decision that has to be made is whether CW or pulsed Doppler instruments should be used. In sections 4.4.1 and 4.4.2 the advantages and limitations of both systems were discussed, but mainly from a technical point of view. In section 10.3 some practical limitations of pulsed Doppler devices were mentioned. In the present section these aspects will be briefly summarized and discussed in relation to the clinical use of both types of instrumentation.

CW Doppler instruments are easy to operate and they are inexpensive. The maximum frequency that can be detected unambiguously with these systems is theoretically unlimited. This is especially important when velocities are recorded within the stenosis to estimate the degree of carotid artery narrowing (see Chapter 3). Velocity recording within a stenosis is rather easy because of the relatively wide ultrasound beam in CW systems. The latter also facilitates velocity imaging of the cervical carotid arteries.

Although in theory the average velocity over the cross-sectional are of an artery can be determined with CW Doppler, at least when the whole cross-sectional area of the blood vessel lies within the ultrasonic beam, accurate assessment of this velocity is still difficult (see Section 4.6). Besides, problems are encountered in processing the CW Doppler signal due to sampling over the full range. Vessel wall motion artefacts can mask the presence of high velocity information, both when a zero-crossing meter and audio spectrum analysis are used (see Section 4.4.1). This is a special problem when high velocity turbulence distal to a stenosis leads to vessel wall vibrations.

A major advantage of pulsed Doppler systems is that velocity information can be obtained from various sites in the vessel along the ultrasonic beam. Hence contamination of the velocity signal by

unwanted signals like those from the vessel wall and veins, can largely be prevented. Information about the angle between the ultrasound
beam and the direction of blood flow can be obtained with pulsed
systems when they are combined with B-mode imaging (see Chapter 9).
Pulsed Doppler devices allow the recording of velocity profiles at
discrete time intervals during the cardiac cycle, at least when multi-channel systems are used (see Section 10.3). Besides, pulsed
systems possess the possibility of determining transcutaneously the
carotid artery diameter and the relative changes in diameter of this
artery during the cardiac cycle. The accuracy of transcutaneous determination of the absolute artery diameter is still limited, but
the relative changes in diameter can be determined rather precisely
(see Section 4.6 and Table 10.1).

Although localization of an artery is relatively easy with multi-
channel pulsed Doppler systems, positioning and maintenance of the
sample volume within a tight stenosis is difficult, especially when
narrow sound beams are used. The major drawback of pulsed devices
is that the maximum velocity that can be detected unambiguously is
limited so that, at the present state of the art, no reliable velocity recordings can be made within tight stenoses. In general pulsed
Doppler systems are more difficult to operate than CW systems.

Although some of the approaches need more detailed clinical evaluation, the techniques presented in the clinical chapters of this
book show that lesions with substantial narrowing of the carotid
arteries (more than 50-60% decrease in diameter) can be detected
rather accurately, both with CW and pulsed Doppler systems. This
indicates that in the detection of these lesions there is no specific
need for pulsed systems. However, at the present state of the art,
no definite conclusions can be made about which of these systems
in detecting stenoses of more than 50% (decrease in diameter) are
most accurate. For most of the techniques using CW Doppler, it is recommended that systems in which forward and backward audio information can be recorded separately should be used so that contamination
of the arterial signal with venous signals can be prevented, allowing
a more precise determination of the diastolic amplitude of the velocity waveform (see Chapter 6).

As mentioned before, early diagnosis of lesions without substantial narrowing of the carotid arteries is important because they are believed to be a source of emboli. There are promising indications that these lesions can be diagnosed both with CW (see Chapter 8) and pulsed Doppler systems (see Chapter 9), although, at the present state of the art, the accuracy of the techniques is limited. In general the diagnosis of lesions with slight degrees of stenosis is focussed on the detection of disturbances in the flow pattern. At present these disturbances are diagnosed by determining the amount of broadening of the audio spectrum. It needs to be investigated whether the on-line recording of velocity profiles at discrete time intervals during the cardiac cycle and instantaneous velocity waveforms at various sites in the vessel along the cross-section may improve the diagnosis of arterial lesions at an early stage of the disease.

10.5.3 Audio Spectrum Analysis

As discussed in section 4.3 and chapter 8 the use of audio spectrum analysis has some major advantages over the use of analog signal processing. In on-line systems, which are commercially available, spectrum analysers yield information about the maximum and mean velocity as well as the velocity distribution of the red blood cells and hence about the flow pattern. The data presented in chapters 4, 7, 8 and 9 allow the conclusion that audio spectrum analysis can be considered to be an asset in the diagnosis of cerebrovascular disease so that its clinical use can be recommended.

It should be noticed that in most of the commercially available spectrum analysers, which are decreasing in price rather rapidly, additional processing is required to obtain the maximum and mean velocity as an instantaneous function of time (see Section 4.3.3).

Whether representation of the frequency intensity in color rather than in grey scale has advantages (see Chapter 7) requires further investigation.

10.5.4 Should One Combine Doppler Systems With An Imaging Technique?

Although in the hands of experts lesions in the cervical carotid arteries can be detected accurately with CW and multi-channel pulsed Doppler systems without imaging these vessels, velocity imaging has been shown to be helpful in localizing the velocity sample site, especially since local examination of these arteries is performed more frequently. Besides, proper localization of the site of recording is useful in follow-up studies, for instance, in evaluating the progress of the disease or the results of surgical or conservative treatment. Velocity imaging may also be useful for educational purposes. Both CW and multi-channel pulsed Doppler systems can easily be combined with velocity imaging, although imaging is generally easier with CW than with pulsed devices because of the wider sound beam in the former device. When single-channel pulsed Doppler systems are used, echo-imaging (e.g. B-mode imaging) is also required to localize the site of sampling.

At the present state of the art, it seems reasonable to combine CW and multi-channel pulsed Doppler instruments with a velocity imaging device for proper velocity localization. Whether velocity imaging can be used to estimate the degree of carotid artery narrowing (see Chapter 8) remains a subject for further investigation.

10.5.5 Real-Time Imaging

Recently some case reports on the use of real-time imaging in the detection of atheromateous disease in the carotid arteries have been published (Hashway and Raines, 1980). Although the first results look promising, further clinical experience is necessary to be able to indicate the utility of these instruments in the diagnosis of lesions in the cervical carotid arteries.

REFERENCES

Anliker, M. (1978). Diagnostic analysis of arterial flow pulses in man. In Cardiovascular system dynamics (Edited by J. Baan, A. Noordergraaf, and J. Raines). pp. 113-123, MIT Press Cambridge.

Barnes, R.W., Bone, G.E., Reinertson, J., Slaymaker, E.E., Hokanson, D.E., and Strandness, D.E. (1976). Non-invasive ultrasonic carotid angiography: prospective validation by contrast arteriography. Surgery 80, 328-335.

Blackshear, W.M., Phillips, D.J., Thiele, B.L., Hirsch, J.H., Chikos, P.M., Marinelli, M.R., Ward, C.J., and Strandness, D.E. (1979). Detection of carotid occlusive disease by ultrasonic imaging and pulsed Doppler spectrum analysis. Surgery 86, 698-706.

Brandestini, M. (1978). Topoflow - a digital full range Doppler velocity meter. IEEE Transactions on Sonics and Ultrasonics SU-25, 287-293.

Eyer, M.K., Brandestini, M.A., Phillips, D.J., and Baker, D.W. (1981). Color digital echo/Doppler image presentation. Ultrasound Med. Biol. 7, 21-31.

Giddens, D.P., Mabon, R.F., and Cassanova, R.A. (1976). Measurements of disordered flows distal to subtotal vascular stenoses in the thoracic aortas of dogs. Circ. Res. 39, 112-119.

Green, P.S., Taenzer, J.C., Ramsey, S.D., Holzemer, J.R., Suarez, J.R., Marich, K.W., Evans, T.C., Sandok, B.A., and Greenleaf, J.F. (1977). A real-time ultrasonic imaging system for carotid arteriography. Ultrasound Med. Biol. 3, 129-142.

Hashway, R. and Raines, J. (1980). Real-time ultrasonic imaging of the peripheral arteries: technique, normal anatomy, and pathology. Cardiovasc. Dis. 7, 257-265.

Hoeks, A.P.G., Reneman, R.S., and Peronneau, P.A. (1981). A multi-gate pulsed Doppler system with serial data processing. IEEE Transactions on Sonics and Ultrasonics SU-28, 242-247.

Hoeks, A.P.G., Ruissen, C.J., and Reneman, R.S. (1980). A multi-gate multi-purpose pulsed Doppler system. Fed. Proc. 39, 1177.

Peronneau, P.A., Bournat, J.P., Bugnon, A., Barbet, A., and Xhaard, M. (1974). Theoretical and practical aspects of pulsed Doppler flowmetry: real-time application to the measure of instantaneous velocity profiles in vitro and in vivo. In Cardiovascular applications of ultrasound (Edited by R.S. Reneman). pp. 66-84, North-Holland/American Elsevier, Amsterdam-London-New York.

Pourcelot, L. (1977). Echo-Doppler systems - Applications for the detection of cardiovascular disorders. In Echocardiology with Doppler applications and real-time imaging (Edited by N. Bom). pp. 245-256, Martinus Nijhoff, The Hague.

Reneman, R.S. What measurements are necessary for adequate evaluation of the peripheral arterial circulation? Cardiovasc. Dis. in press.

Sandmann, W., Peronneau, P.A., Schweins, G., Bournat, J., and Hinglais, J. (1978). Turbulenzmessung mit dem Doppler-Ultraschallverfahren: Eine neue Methode der Qualitätskontrolle in der Arterienchirurgie. In Ultraschall-Doppler-Diagnostik in der Angiologie (Edited by A. Kriesmann and A. Bollinger). pp. 77-81, Georg Thieme Verlag, Stuttgart.

Wille, S.Ø. (1979). Numerical models of arterial blood flow. Thesis. Institute of Informatics. University of Oslo.

Index of Subjects